NEHEMIAH GREW M.D., F.R.S.
A Study and Bibliography
of his Writings

Nehemiah Grew

NEHEMIAH GREW M.D., F.R.S.

*A Study and Bibliography
of his Writings*

by
WILLIAM LeFANU F.S.A.

ST PAUL'S BIBLIOGRAPHIES
WINCHESTER

OMNIGRAPHICS INC.
DETROIT
1990

First published 1990 by
St. Paul's Bibliographies
West End House
1 Step Terrace
Winchester
Hampshire   SO22 5BW
England

Distributed in North and South America by
Omnigraphics Inc., Penobscot Building, Detroit

British Library Cataloguing in Publication Data
Le Fanu, W. R. (William Richard, *1904*–)
Nehemiah Grew.
1. Botany.   Grew, Nehemiah 1641–1712
I. Title
581'.092'4

ISBN 0–906795–43–5

Library of Congress Catalog Card 89–43507

THIS EDITION IS LIMITED TO 500 COPIES

Typeset in Monophoto Lasercomp Bembo by
August Filmsetting, Haydock, St. Helens
Printed on long-life paper ∞ in England by
St. Edmundsbury Press, Bury St. Edmunds

for Elizabeth

Contents

Prefatory Note   xi

Acknowledgments   xiii

References   xv

Illustrations   xvii

GREW'S LIFE AND CAREER: A SUMMARY NOTE   I

THE WRITINGS OF NEHEMIAH GREW

  I   Disputatio de Liquore Nervoso 1671   5

 II   Anatomy of Plants 1672–1682:   7

      IIA   The Anatomy of Vegetables begun
             1672   8

      IIB   An Idea of a Phytological History—
             The Anatomy and Vegetation of
             Roots   1673   14

      IIC   A Discourse concerning Mixture 1675   17

      IID   The Comparative Anatomy of Trunks
             1675   18

      IIE   Experiments in Consort of Luctation
             1678   20

      IIF   The Anatomy of Plants 1682   23

 III Musaeum Regalis Societatis 1681:   28

      IIIA   The Royal Society's Museum and its
             Exhibits   32

      IIIB   Musaeum 1681—Comparative
             Anatomy of Stomachs and Guts   40

IV	Sea-water Made Fresh 1683		44
V	Bitter Purging Salt 1695		49
VI	Cosmologia Sacra 1701		53
VII	The Royal Society:		
	VIIA	Grew as Secretary and Editor 1677–79	55
	VIIB	Contributions to *Philosophical Transactions*:	
		1 Observations touching the Nature of Snow 1673	60
		2 The Pores in the Skin of the Hands and Feet 1684	61
		3 Observationes de Morboso Liene 1691	62
		4 The Food of the Humming Bird 1693	62
		5 The Number of Acres in England 1711	64
VIII	Enquiries relating to New England and the Indians 1690		67
IX	Account of Henry Sampson 1705		68
X	Correspondence		70
XI	Funeral Sermon by John Shower 1712		72
THE BIBLIOGRAPHY			75
I	De Liquore Nervoso 1671		76
II	The Anatomy of Plants:		
	IIA	The Anatomy of Vegetables Begun 1672	77
	IIB	An Idea of a Phytological History 1673	85
	IIC	A Discourse concerning Mixture 1675	88
	IID	The Comparative Anatomy of Trunks 1675	90
	IIE	Experiments in Consort of Luctation 1678	93
	IIF	The Anatomy of Plants 1682	97
	IIG	Anatomy of Plants—Related Papers	106
III	Musaeum Regalis Societatis 1681		109
IV	Sea-water Made Fresh		122
	IVA	Sea-water Made Fresh 1683	124
	IVB	Fitzgerald's Pamphlets	127
	IVC	Fitzgerald's Supplementary Pamphlets	130

	IVD	Walcot's Pamphlets	134
V		Bitter Purging Salt 1695	135
VI		Cosmologia Sacra 1701	143
VII		Contributions to *Philosophical Transactions*:	
	1	The Nature of Snow 1673	147
	2	The Pores in the Skin of Hands and Feet 1684	147
	3	De Morboso Liene 1691	147
	4	The Food of the Humming Bird 1693	148
	5	The Number of Acres in England 1711	148
VIII		Enquiries relating to New England and the Indians 1690	150
IX		Account of Henry Sampson 1705	151
X		Correspondence	152
XI		A Funeral Sermon upon the Death of Dr Grew 1712	154

APPENDIX	I	Donors to the Repository, and Subscribers to *Musaeum* or *The Anatomy of Plants*	155
APPENDIX	II	*Philosophical Transactions*— six issues edited by Grew 1678–79	163
APPENDIX	III	'Papers of Dr N. Grew' [*c.* 1709]	167

GENERAL INDEX	169
INDEX OF LIBRARIES	180

Prefatory Note

NEHEMIAH GREW was a pioneer in exploring the physiology of plants and the comparative anatomy of animals. He was an active Fellow of the Royal Society and one of its busiest officers in its first great period during the 1670s. Born in Warwickshire into a dissenting family in 1641, when the Civil War was beginning, he was educated at Cambridge, and practised as a physician in the City of London, where he died in 1712 still a devout Presbyterian.

My study of his work surveys writings on many subjects, and primarily his publications in botany. It has been made with the support of the Royal Society, to whom I am profoundly grateful for a generous grant towards the expense of my research and for the privilege of using their Library and Archives.

WILLIAM LeFANU
Shottesbrook, Boreham,
Chelmsford, 1989

Acknowledgments

MANY INSTITUTIONS, librarians and other friends have given me information about Nehemiah Grew. I return cordial thanks to the Royal Society, the Royal College of Surgeons of England, the British Library Board, the Trustees of the British Museum (Natural History), the House of Lords Record Office, the Linnean Society, the Royal College of Physicians, the Royal Pharmaceutical Society, the Royal Society of Medicine, the Society of Antiquaries, the University of London Library, the Wellcome Institute. Cambridge: Trinity College and Magdalene College (Pepysian Library). Dublin: Trinity College, the Royal College of Physicians of Ireland, the Royal College of Surgeons of Ireland. Edinburgh: the National Library of Scotland, the Royal Botanic Garden, the Royal College of Physicians of Edinburgh. The Hertfordshire Record Office. Leeds University Brotherton Library. Oxford: the Bodleian Library, Christ Church, Magdalen College. Paris: Académie nationale de Médecine, Bibliothèque interuniversitaire de Médecine, Bibliothèque Nationale, Muséum national d'Histoire naturelle. The Hague: Royal Library, Central Catalogue. Leiden State University. Montreal: McGill University, Osler Library. Baltimore, Maryland: Institute of the History of Medicine. Boston and Cambridge, Massachusetts: Harvard University, Houghton Library and Countway Library of Medicine. New Haven, Connecticut: Yale Medical Historical Library. Philadelphia, Pennsylvania: American Philosophical Society, College of Physicians of Philadelphia, Library Company of Philadelphia. Washington, DC (Bethesda, Maryland): National Library of Medicine.

Personal thanks are due to Mr E. Alting, Miss Jean Archibald, Miss J.G. Beevers, Dr Whitfield, J. Bell, Jr., Dr John B. Blake, Mr R. Breugelmans, Mme P. Casseyre, Mr Dennis Cole, Mr Eustace Cornelius, the late Dr George Corner, Mr Timothy J. Christ, Mrs Judith Diment, Mr Robert Donaldson, Miss Sheila Edward, Miss Joan Ferguson, Dr Philip Gaskell,

Mme Y. Gueniot, Mr G.C. Harries, M.C. Hustache, Mr D.J. Johnson, Dr Richard Luckett, Miss M.F. MacKenzie, Mr David McKitterick, Mr J.F.A. Mason, M.B. Molitor, Mr P.S. Morrish, Mr Elliott H. Morse, Mr Leonard Payne, Mr Robin Price, Mr Julian Roberts, Mr N.H. Robinson, Mr L. Rowlands, Mr David Stewart, Ms R. Stephen, Dr C.O. von Sydow, Mrs V.I. Tuttle, M.G. Velay, Dr Edwin Wolf 2d.

I have received invaluable encouragement and advice from Dr Michael Hunter, the late Sir Geoffrey Keynes, Sir William Paton, Dr Alex Sakula, and Mr Alwyne Wheeler. Mr Ian F. Lyle has made frequent research for me and provided me with transcripts and photographs. Mrs Marion May has admirably retyped my rough drafts. Mr David Way has prepared my text for the printer and designed the book's form. Mr Robert Cross has been a most considerate and helpful publisher through my long delays.

References

H.B. ADELMANN (1966) *Marcello Malpighi and the Evolution of Embryology*
——— (1975) *The Correspondence of Marcello Malpighi*

AGNES ARBER (1953) 'From medieval herbalism to modern
botany' *Science Medicine and History* 1: 317–36,
and earlier papers

THOMAS BIRCH (1756) *The History of the Royal Society*

JEANNE BOLAM (1973) 'The botanical Works of Nehemiah Grew'
Notes & Records of the Royal Society 27: 219–23

G.S. BOULGER (1890) 'Grew, Nehemiah' *Dictionary of National
Biography* 23: 166–68

RENATE BURGESS (1973) *Portraits of Doctors and Scientists, a Catalogue.*
Wellcome Institute

WILLIAM CARRUTHERS (1902) 'The Life and Work of Nehemiah
Grew' *Journal of the Royal Microscopical
Society* 129: 129–41

F.J. COLE (1944) *A History of Comparative Anatomy*
——— (1969) *The Cole Library of early Medicine and Zoology* [Reading]:
Catalogue by N.B. Eales

JOHN F. FULTON (1961) *A Bibliography of the Honourable Robert Boyle,*
2nd Edition

A.R. & M.B. HALL (1965–) *The Correspondence of Henry Oldenburg*

BLANCHE HENREY (1975) *British Botanical and Horticultural Literature before
1800*

REFERENCES

MICHAEL HUNTER (1981)   *Science and Society in Restoration England*
—— (1982a)   'Early Problems in professionalising Scientific
          Research' *Notes and Records of the Royal Society* 36:
          189–209
—— (1982b)   *The Royal Society and its Fellows 1660–1700*
—— (1985)   'The Cabinet institutionalized; the Royal Society's
          Repository and its Background', in O. Miller and A.
          Macgregor *The Origins of Museums.*

R.W. INNES SMITH (1932)   *English-speaking Students of Medicine at the
          University of Leiden*

C.R. METCALFE (1972)   'Grew, Nehemiah' *Dictionary of Scientific
          Biography* 5: 534–36

K.F. RUSSELL (1963)   *British Anatomy 1525–1800, a Bibliography* [2nd
          edition 1987]

ALEX SAKULA (1982)   'The Waters of Epsom Spa' *Journal of the Royal
          College of Physicians of London* 16: 124–28
—— (1984)   'Dr Nehemiah Grew and the Epsom Salts' *Clio Medica*
          19: 1–21

BENNET WOODCROFT (1854) *Alphabetic Index of Patentees of Inventions.*
Reissue with additions by D.R. Jamieson 1969

Several of the authorities listed above mention earlier notices of Grew,
who is naturally discussed also in the histories of science. I have not
listed here authors quoted in full in the text on single special points.

Illustrations

Frontispiece Nehemiah Grew. Portrait by Robert White, from
              Cosmologia Sacra, 1701
Plate 1 *Musaeum Regalis Societatis*
         Plate 6: Skull, Eggs and Nests of Birds   29
      2 *Comparative Anatomy of Stomachs and Guts*
         Musaeum Plate 30: Gullets of Birds   31
      3 *The Anatomy of Plants*
         Plate 1: The Seed in its Vegetation   101
      4 *The Anatomy of Plants*
         Plate 5: The Generation of Roots   102
      5 *The Anatomy of Plants*
         Plate 18: The Trunks of Plants cut transversely   103
      6 *The Anatomy of Plants*
         Plate 51: Air Vessels in a Vine-leaf   104
      7 *Daniel Colwall* F.R.S.
         Portrait by Robert White, from *Musaeum* 1685   119

The Frontispiece and Plates 1–2, 4–7 are reproduced by permission of
the Royal College of Surgeons of England; Plate 3 by permission of the
Wellcome Institute Library, London.

Grew's Life And Career:
a summary note

NEHEMIAH GREW was born in September 1641 at Mancetter in Warwickshire, a village twelve miles north of Coventry and a little south of Atherton where his father Obadiah Grew had come from Balliol College, Oxford in 1632 to be Master of the School. In the winter of 1636–37 Obadiah married Helen Sampson, a widow with two sons; he was appointed Vicar of St Michael's Coventry in 1645 and admitted a Doctor of Divinity at Oxford in 1651. But he was extruded from his living as a dissenter, when the Act of Uniformity came into force in the summer of 1662, a few months after Nehemiah graduated at Cambridge.

Mrs Grew's first husband William Sampson was steward in the Puritan household of Sir Henry Willoughby, baronet, of Risley in Derbyshire, and had published plays and poems; he died in 1636 leaving two sons Henry and William, both of whom became Fellows of Pembroke College, Cambridge. Nehemiah Grew followed them to Pembroke, entering as a Pensioner on 21 June 1658, when he was seventeen, and graduated B.A. in 1662, but the Act of Uniformity, which extruded his father, similarly barred him as a nonconformist from taking any further degree. He went home to Coventry and worked at botany, chemistry and anatomy during the next decade. His teachers are not known, though he may have studied natural science at Leiden, where an Englishman 'Nehemiah Weit, aged 20' enrolled in the Faculty of Philosophy (Natural Science) in 1663; Innes Smith (1932) suggested that 'Weit', otherwise untraced, was a mistake for 'Grew'.

Henry Sampson, Grew's half-brother and life-long friend, was ejected from his Fellowship at Pembroke and his Rectory at Framlingham in Suffolk when the Church of England was re-established in 1662, and was prevented by the Conventicle Act of 1664 from continuing his private ministry there. So he travelled in France, Switzerland and Italy studying anatomy and medicine, and graduated M.D. at Leiden in 1668. Thereafter he practised as a physician in London. Following his example Grew obtained the Leiden M.D. in 1671 and began to practise medicine at Coventry.

Henry Sampson showed Grew's first botanical essay in 1670 to Henry Oldenburg, the Secretary of the Royal Society, who asked Bishop John Wilkins's advice about it. Wilkins was a founder Fellow and a most active member of the Society's Council, which recommended publication early in 1671. Grew was elected a Fellow in November, and his essay *The Anatomy of Vegetables Begun* was printed in December. These dates are in Grew's preface to his *Anatomy of Plants* 1682, with others recorded in Birch's *History of the Royal Society* 1756.

At Wilkins's instance Grew was invited to London and appointed in April 1672 the Society's 'Curator for the Anatomy of Plants' for one year at a salary of £50 to be subscribed by Fellows; this post entailed his showing regular demonstrations at the Society's meetings. After the first year Grew wrote to Oldenburg asking that his employment might 'be continued for at least five years more' with £250 'payd me now altogether' or by £50 a year; 'this my Lord Brouncker [the President] hath already engaged for', but Grew asked Oldenburg to request the Council of the Society 'to approve security for the full and quarterly payment of the moneys'. The autograph (Royal Society: M.M.4,f.71) is undated, but of late 1672, for Grew wrote that he had been employed by the Society 'for a year and upward'. This letter was not seen by Professor and Mrs Hall when they edited Oldenburg's *Correspondence*, as it had been filed in the 'Miscellaneous Manuscripts', where it was discovered by Dr Michael Hunter who published it in *Notes and Records of the Royal Society* 36 (1982), 193–95.

During the next ten years Grew gave most of his time to the Society: he lectured regularly between 1672 and 1679, became joint Secretary with Robert Hooke after Henry Oldenburg's death on 5 September 1677 till the end of 1679, edited the *Philosophical Transactions* through 1678, published his Catalogue of the Society's *Musaeum* in 1681, and in 1682 gathered his lectures into *The Anatomy of Plants*, a landmark in the literature of botany. It is tragic to recall that in the same year, 1682, his learned and pious father, aged 75, was sent to prison, under the Five Mile Act of 1665, for teaching at the Presbyterian Academy in Coventry (I. Parker *Dissenting Academies* 1914, p.138).

Grew's age was forty-one when he published *The Anatomy of Plants*. He completed no more botanical work, but his later manuscripts and his collection of exotic seeds, both acquired by Hans Sloane, show that he retained his interest. He had been elected an Honorary Fellow of the College of Physicians in 1680, when it opened its doors to medical graduates from beyond Oxford and Cambridge, and through the following thirty years devoted himself to his practice in the City of London. Between 1684 and 1711 he published pamphlets and papers on a wide variety of subjects and one large philosophical book.

Nehemiah Grew died suddenly at the age of seventy in 1712, still active in practice. He was twice married and was survived by his second wife Elizabeth Dodson of Cheshunt, with their son and two daughters. (Hertfordshire Record Office: Cheshunt Parish Records D/P 29).

The writings of Nehemiah Grew

I DISPUTATIO DE LIQUORE NERVOSO 1671

Grew's first printed work was his Latin dissertation for the Doctorate of Medicine at Leiden University. He enrolled as a Candidate in the Faculty of Medicine on 6 July 1671, describing himself for the Register as 'Warwicensis, aet. 30' (Innes Smith, p.102). A week later he was admitted Doctor of Medicine on 14 July, when his Disputation on the Nervous Liquor was printed by the Widow and Heirs of John Elzevir, printers to the University. Grew dedicated it to his father, his half-brother Henry Sampson, and his friend Abraham Clifford. Grew read this Dissertation again in 1675 before the Royal Society; a manuscript copy, in a hand more careful than Grew's usual writing, is in the Society's archives.

Abraham Clifford (1628–75) was, like Grew, a Warwickshire man and, like Sampson, was ejected as a Presbyterian from his Fellowship of Pembroke College, Cambridge and his Rectory at Quendon, Essex in 1662; he had graduated at Cambridge B.A. 1650 and B.D. 1660, after his ejection he graduated M.D. at Oxford 1670 and practised in London, but died in 1675 aged forty-seven (Venn, *Alumni Cantabrigienses*). Grew named only one authority in his Disputation: 'Dr F. Sylvius, Professor of the Practice of Medicine in this University', that is Franz Deleboë or Dubois, usually called Sylvius. The presiding Rector of the University was Abraham Heidan D.D., Professor Ordinarius in the Faculty of Theology and Pastor of the Church.

The Disputation comprises twenty-five 'theses' discussing the origin and constitution of the 'nervous liquor' generally believed to convey impulses through the nerves, which were conceived as hollow and analogous to the blood-vessels. The main argument is a

5

theoretical analysis of the blood, determining from which of its constituents this liquor derives; 'for by what other instrument', Grew asked in Thesis XV, 'can the mental and bodily movements be accomplished?'

Thesis XXI suggested that the watery and oily parts of the nervous liquor descend slowly into the nerves appropriate to the sense organs, while the other parts of the liquor are thrown into the lower and harder nerves. The final Thesis XXV predicted that 'discovery and explanation of the cause of thought, sleep, dreams, movement of the heart, the system of all organically moving parts, nutrition, manner of generation of the various humors, madness, melancholy, lethargy, apoplexy, epilepsy, may not perhaps be so difficult as they have been for so long'.

II   ANATOMY OF PLANTS   1672–1682

Grew began his research into the development of plants at Coventry about 1664. Through the support of the Royal Society he extended this pioneering work and made it known by lectures and small books during the 1670s, finally publishing all his discoveries in 1682.

John Wilkins, Bishop of Chester, who promoted the Society's publication of Grew's first botanical book during 1671, proposed that he should be appointed Curator for the Anatomy of Plants for a year at a salary of £50 to be subscribed by Fellows. When the Society's Council approved this appointment on 18 April 1672 'a form was ready for the subscription of such as would freely contribute to so good a Work'. Grew who was present on 8 May promised to speak about plants at the next meeting day, and Robert Hooke, who had been Curator of Experiments since 1662, was desired 'to deliver to him the Society's microscope'. Grew duly gave his first demonstration on 22 May 1672, when he 'shewed the Company through a Microscope the conformation of the Pith in Vegetables, and gave in a larger account in writing'.

His appointment was continued from 1673 for five years more, since he claimed that he had foregone opportunity of advancement and profit when he gave up medical practice at Coventry to work for the Society in London. During those five years he arranged demonstrations and lectures at the regular meetings of the Society, exhibiting specimens, drawings and experiments on botany, chemistry and the comparative anatomy of animals. Birch in his *History of the Royal Society* recorded occasional comments by Grew at meetings of the Society, for instance on 4 February 1674/5 after Hooke's observations on human muscle.

A few of Grew's Discourses were published separately: three small books on botany and two on chemistry. All his Discourses were collected to form his most original and influential book *The Anatomy of Plants*, in 1682, except his lectures of 1677 on *The Comparative Anatomy of Stomachs and Guts*, which he had 'subjoined' to his Catalogue of the Royal Society's Museum in 1681. Though that work, *Musaeum*, was published a year before *The Anatomy of Plants*

I defer description of it till after my account of the botany and chemistry books. He published no more on these subjects, but a gathering of 'Dr Grew's botanical Papers' in the British Library forms Sloane MS 2145 and includes letters of 1709 and 1710, showing that he was actively interested in botany till near the end of his life.

IIA   THE ANATOMY OF VEGETABLES
BEGUN   1672

Nehemiah Grew and Marcello Malpighi, working independently and far apart, reported their botanical discoveries to the Royal Society at almost the same time, in November and December 1671. They inaugurated a new science, the anatomy and physiology of plants; no recent botanist had moved as far beyond description and taxonomy.

In antiquity Theophrastus (371–287 BC) wrote a similar treatise *De Causis...*, as well as his taxonomic *De Historia Plantarum*; there are Renaissance editions and Latin translations of both works. Grew was a good scholar but made no reference to Greek science, perhaps obeying the Royal Society's motto *Nullius in verba*; his half-brother Henry Sampson, who influenced him so profoundly, had been, before qualifying in medicine, a lecturer on Greek philosophy at Pembroke College, Cambridge, where Grew was an undergraduate in 1658–61. During those years John Ray was an active Fellow and tutor at Trinity; he published in 1660 his pioneer county flora of Cambridgeshire, in which he quoted from Theophrastus's two botanical treatises. Grew was not taught by Ray, perhaps not even directly influenced by him towards botany; but in *An Idea of a Phytological History* (1673) he praised 'our learned countryman Mr Ray for his very laudable pains in adjusting the order and kindred' of plants. They had little personal contact in subsequent years, while working in parallel and both publishing through the Royal Society. (C.E. Raven *John Ray naturalist*, 2nd ed. Cambridge 1950, reissued 1986, chapter viii, pp.180–200 'The structure and classification of plants', and pp.200–201 'Note. Ray's relations with Nehemiah

Grew'). Ray's *Flora of Cambridgeshire* was translated and edited by A.H. Ewen and C.T. Prime, and published by Wheldon & Wesley of Hitchin in 1975.

Howard B. Adelmann in his great edition of *Malpighi* (Cornell 1966, pp.1095 ff.) discussed the botanists who anticipated some of the discoveries of Malpighi and Grew, naming particularly Giovanni Aromatari, who published his observation of the development of the seed from the pollen – *De generatione plantarum ex seminibus* – at Venice 1625, and Honoré Fabri, whose treatise *De Plantis* appeared at Rome 1666.

Grew recorded his work at first in Latin: an autograph draft *Anatomia Vegetalium Inchoata* survives among his papers, but he changed to English. 'Before I had ventured very far', he wrote when publishing his completed work, 'in the year 1668 I imparted my design to my brother-in-law (i.e. half-brother) the learned Dr Henry Sampson, who excited me to a vigorous and accurate persecution of it, mentioning a pertinent passage of Dr Glisson. After I had finished my first book I put some part of it into the same hand, who in the year 1670 communicated the same to Mr Oldenburg then Secretary of the Royal Society'. Henry Oldenburg referred the paper to John Wilkins, at whose suggestion Grew sent his whole draft to the Royal Society; on 11 May 1671 the Society's Council ordered its printing.

Grew signed his Epistle Dedicatory to Bishop Wilkins on 10 June, writing that many of his observations 'have lain dormant near seven years'. In July he went to Leiden where he received his medical degree, and soon began to practise as a physician at home in Coventry. While he was away three plates for the book were engraved, with the individual figures in disarray. The printing order was confirmed by the Royal Society only on 9 November, and on 15 November Grew was elected into Fellowship. The book was printed quickly: four copies were presented by the Society's printer Spencer Hickman at the meeting on 7 December 1671: one for the Society, which still possesses it, one for the President, Lord Brouncker, and two for the Secretaries, Henry Oldenburg and Thomas Henshaw, secretary from 1669 till succeeded by John Evelyn at the annual election on 30 November 1672 (Evelyn's

Diary: 'I was chosen Secretary'.)

'Not long afterwards', Grew recalled, 'I received news from London that the same day in which my book was presented there was also presented a manuscript (without figures) from Signor Malpighi upon the same subject, dated at Bologna November 1st 1671'. (Dates and quotations from *Anatomy of Plants* 1682, Preface, and the half-title there of *Anatomy of Plants Begun*, second edition.)

Malpighi was thirteen years older than Grew, and already at forty-three well known as physician and anatomist. Many botanists have compared their researches: Malpighi's work was more profound, but Grew's equally original and published first, if brief priority has significance. Grew decided to write no more on botany in the face of Malpighi's achievement, but the Royal Society encouraged them both to continue. Grew reported his work piecemeal to the Society in a series of Discourses, five of which were published individually before the whole series was gathered in *The Anatomy of Plants* 1682; meanwhile the Society had published Malpighi's Latin account of his researches in two volumes: *Anatome Plantarum* I, 1675 and II, 1679. Howard B. Adelmann in his biography of Malpighi wrote 'nothing but courtesy, mutual respect and generous appreciation of one another's efforts marked their relations in the years that passed till the completion of their work'; he published all Malpighi's *Correspondence* in 1975.

The Anatomy of Vegetables Begun was printed by 7 December 1671, though dated 1672 in the imprint; it was recorded in the Term Catalogue of 7 February 1671/2. Nearly a year after its publication Grew's conclusions were criticised by Martin Lister in a long letter to Oldenburg dated 30 November 1672; on 11 December Grew wrote a defence covering thirty-two particular points, which Oldenburg transcribed for Lister. (Royal Society archives and Bodleian Library; texts in Hall *Oldenburg's Correspondence*.)

For the 'Second Edition', which forms 'The First Book' in the collective *Anatomy of Plants*, Grew revised his text very slightly but had new plates engraved displaying more figures. A French translation, made by Louis Le Vasseur with Grew's help, was published in Paris in 1675, with a second edition in 1679; it was reprinted at Leiden with other works, including Mesmin's version of Grew's

Luctation, in 1685 and reissued in 1691; this collection was published in Italian at Venice in 1763. A Latin version, which Grew considered 'unskilful', was included in the *Miscellanea Curiosa* of the German Academia Naturae Curiosorum in 1678.

The French editions have sometimes been confused. They were:

1675   *Anatomy of Vegetables Begun*, tr. L. LeVasseur. Paris, L. Roulland.

1679   The same reissued: Paris, A. Dezallier.

1679   *Luctation*, tr. G. Mesmin. Paris, E. Michallet.

1685   *Anatomy of Vegetables begun* (LeVasseur); N. Dedu *L'Ame des Plantes* [from: Paris, E. Michallet 1682]; *Luctation* (Mesmin); Boyle *Tastes and Odours* [from: Oxford 1675]. Leiden, P. vander Aa.

1691   The same collection reissued by vander Aa.

1698   *Luctation* (Mesmin). Paris, B. Girin.

The Anatomy of Vegetables Begun is the first published study of the development of a plant from seed to seed, briefly surveying the succession of seed, root, stem, branch, leaf, flower, fruit and the new seed. All these Grew described more fully in his demonstrations during the next six years. Defining his subject as 'the visible constitution and uses of the several parts of plants' he recorded three aspects of his work: accurate detail of dissections, discussion of his concept of the mechanisms of development, and a teleological theory of the various uses of a plant: for human life, in God's scheme of creation, and for the plant itself. To a great extent he invented his own terminology, usually familiar English words in specialised senses. Describing his dissection of a bean-seed in Chapter I, he named the 'radicle' which becomes the root, the 'plume' which becomes the stem, and the 'lobes' (cotyledons) producing the 'dissimilar leaves'; cells he called 'bubbles' in 1672, but changed to 'cells or bladders' in 1682. The term 'cell' was introduced by Hooke in 1665 (*Micrographia*, p.113) and adopted by Grew in 1682 (*The Anatomy of Plants*, p.64). His terms were in general not taken on, yet Johnson in his *Dictionary* (1755) quoted Grew as authority for botanical usage of many words. Grew liked homely analogies, such as comparing the inner coat of bean-seeds to 'wafers under maquaroons'.

The account of sap in Chapters 2 and 3 reported new knowledge; Chapter 5 gave the first analysis of the flower as 'empalement' [calyx], 'foliation' [corolla], and 'attire', [stamens]. The term 'attire' may have been used botanically before Grew's time: Milton in *Lycidas* (1637) called the woodbine (*Lonicera periclymenum*), with its conspicuous anthers, 'well-attired'. Grew carefully described the 'semets' and their 'powder' [anthers and pollen] and recognised the sexual nature of plant reproduction, but was not prepared to publish his belief; he considered the uses of the attire: 'for us, ornament – merely beauty – and distinction [taxonomic variation]; for other animals, food, but what may be the primary and private use of the attire, I now determine not'. Grew declared his knowledge of sex in plants in his Discourse of Flowers on 9 November 1676, but did not publish it till 1682.

In all his observations Grew constantly compared different plants through a long series of trees and herbs. He often drew analogies with animal structures and physiology, comparing the growth of the stem with that of bones and feathers, the movement of sap with animal motion, and the tissue of seeds with the blood-vessels. Most of his work was done with 'the bare eye', while he mentions occasionally using 'indifferent', 'good' or 'very good' spectacle glasses, but 'the use of a microscope, except in some few particulars, I have purposely omitted'.

Grew named three predecessors: Sir Thomas Browne who described the arrangement of florets in compositae (*The Garden of Cyrus* 1658, in *The Works*... edited by Geoffrey Keynes I(1964), 193–225 especially chapters 3–5); Robert Sharrock, who recorded his observations on cotyledons, the development of branches, and the use of thorns (*The History of the Propagation and Improvement of Vegetables*, Oxford 1660, p.36 cotyledons, pp.116, 121, 142 branches, p.140 thorns); Robert Hooke, who described and illustrated the microscopic structure of pores in wood and pith, globulets in leaves, and the diversity of hairs (*Micrographia* 1665, Observations 16–21 and 23–31, with Schemata X–XX).

Alterations in the edition of 1682 from the text of 1672 were mainly trivial: The Imprimatur and Preface were omitted; the Dedication 'To the President of the Royal Society' now named

Lord Brouncker, who had retired in 1677; the Epistle to the Bishop of Chester, John Wilkins, undated in the first edition, was now dated 'Coventry. June 10.1671'; and the reference to Glisson was transferred to the general Preface of the volume.

In the title of this 'First Book' (1682) Grew replaced 'Vegetables' by 'Plants' and in the sub-title 'A general Account of Vegetation founded thereon' became 'Grounded thereupon'. More significantly a few technical terms were changed: Germen became Bud, Seminie and Florie became Seminiform and Florid Attire; Trunk at the end of Chapter III was replaced by Tree. The chapter headings were slightly revised, the paragraphs were numbered in each chapter, and there was inessential rephrasing throughout the text. More important was the addition of a few sentences and paragraphs: Chapter I para. 5 was enlarged and paras 6–7 rewritten in a fuller account of the foramen in the outer coat of certain seeds; para. 13 on the acrospire 'the part which becomes the trunk' was rewritten; para. 25 on dissecting a bean to show the 'seminal root' was prolonged; para. 33 on the radicle being 'suppled' by the inner coat was rewritten. Chapter II para. 30 was rewritten, suggesting that the sap circulates, because it goes down the root and up the trunk. Chapter III, Appendix: para. 5 was added on 'the retrograde motion about every third circle' in the tendrils of Bryony, which seems the first considered notice of the functioning of tendrils.

French translation   1675

In the preface to *The Anatomy of Plants* 1682 Grew wrote 'The First Book a little after it came forth was translated into the French Tongue by Mons. LeVasseur an Ingenious Gentleman in Paris, elegantly and in the Judgment of those who are skilled in that Language with much exactness as to the sense. He having taken special care to have all the difficulties of our own by Me cleared to him'.

Louis LeVasseur in his *Avis au Lecteur* defended his use of new French terms and colloquialisms, and wrote that he took synonyms from the 'Enchyridion Botanicum du sieur Robert Louvell Anglois qui est fort exact'. Lovell's *Panbotanologia sive Enchiridion Botanicum, or a compleat Herball,* Oxford 1659, 2.ed. 1665, includes an 'Index of Latin names with those that are synonymous'. Lovell practised as a physician at Coventry, where Grew lived between 1661 and 1671. Grew sent LeVasseur revisions of his text in several letters, and one of 19/26 November 1674 was printed in its original Latin in Le Vasseur's book; Grew's draft of this letter is in the British Library (Sloane MS 1926,f.198v), with several pages of 'alterations, new interpretations, and a few deletions' all in Latin. Grew wrote that he sends corrections of printer's errors in English, the rest in Latin, and hopes to read them all in French; he ends with greeting to Le Vasseur's brother, and signs 'amantissimus'.

Grew wrote in his Preface of 1682 'In a late book entituled *Philosophia vetus et nova* printed at Noriberg 1682 the learned Author seems to have made use of this translation, for all that he hath taken notice of in that my First Book'. This was probably Jean-Baptiste Du Hamel's book first published in Paris 1663, several times reissued there and at Nuremberg 1681 as 'De consensu veteris et novae Philosophiae' in his *Opera Philosophicorum* (I have used the copy at Cambridge University Library); it was reissued in London 1685 (Wing d2499). There is no direct reference to Grew's work, but brief mention of the study of plants for medical use (Vol.2, 1681, Book 1 'De corporum affectionibus', chap.2, 3, pp.15–16). Du Hamel (1623–1706) was Secretary of the Académie royale des Sciences 1666–97; he was in England with a diplomatic mission in 1669, when he met Boyle and Oldenburg. Louis LeVasseur, M.D., of Montpellier, practised in Paris till 1682; then, as a Huguenot refugee, he obtained denization in England in January 1683 for himself and his wife with their two sons and two daughters (*Huguenot Society Publications* 18 (1911), 159); he was admitted L.R.C.P. on 2 October 1683.

IIb   AN IDEA OF A PHYTOLOGICAL
HISTORY / THE ANATOMY AND VEGETATION
OF ROOTS   1673

Grew's second botanical book combines two series of Discourses which he read to the Royal Society during 1672 and early in 1673:

An Idea of a Phytological History (setting out the aims of his research), with *The Anatomy of Roots* and *The Vegetation of Roots*. It was printed by John Martyn for Richard Chiswell and recorded in the Term Catalogue of 16 June 1673.

As with his previous book, Grew wrote first in Latin: his botanical notes among the Sloane manuscripts include a much-corrected draft of *An Idea* and a fair text of the first part of *The Anatomy of Roots*, both in Latin. A Latin translation made in Germany, which Grew disparaged for its incorrectness, was printed in *Miscellanea Curiosa*, volume 9–10 for 1678–79 published in 1680.

For the collected *Anatomy of Plants* 1682 Grew made a logical separation of these Discourses, using *An Idea* as a general introduction before *The Anatomy of Plants Begun* which he called 'The First Book', while the two parts of *The Anatomy of Roots* became 'The Second Book'. In the title of *An Idea* he now wrote 'Philosophical History' in place of 'Phytological . . .', showing that this Discourse is a survey of the 'knowable' in the 'unknown' parts of botany, not the story of what had already been discovered. He summarised the work of earlier and contemporary botanists in propagating new varieties and introducing exotic plants 'especially natives of the Indies', in classification and nomenclature, and in expounding the use of plants in daily life; but they had studied only 'the parts above ground'.

An Idea is a programme for 'improvements' in knowledge of plant life. Grew hoped to replace 'descriptions yet to be perfected' and 'names not well given'. 'For their Figures it were much to be wished that they were all drawn by one Scale, or at most by Two; one for Trees and Shrubs and another for Herbs'; 'many of their Ranks and Affinities are yet undetermined', though he praised John Ray and Robert Morison for 'adjusting their Order and Kindred'; but 'for the Reason of Vegetation and the Causes of Varieties almost all Men seemed to be unconcerned'. His research was aimed to show how plants grow 'from Seed to Root and Trunk, then all the other parts, to Seed again; how their Aliment is prepared, conveyed, and assimilated; their Sizes, their Shapes as of Roots, of Trunks, of Leaves, and so for the other Parts; their Motions, their Seasons for Spring or Birth, Full Growth, and Teeming; the Period of their

Lives, Annual, Biennial, Perennial'. In his habit of drawing analogies between plant and animal life, which he admitted was excessive, he looked for 'the Oeconomy of the Whole' in their 'feeding, housing, cloathing or protection, and care for themselves and their Posterity'.

Grew proposed 'Five means of inquiry': (1) the proper places and climates for different plants; (2) their anatomy; (3) their contained fluids, with 'applicable experiments', mainly chemical, distinguishing their colours, smells and tastes; (4) their principles whether acid or salt, their substance, whether pithy or lignous, etc.; (5) the external materials – earth, water, air, sun – from which their principles arise. Beyond these five he hinted that it would become possible to discover how these principles unite to form a Vegetable Body, but 'I shall not now conjecture'.

A vast edifice of knowledge has been raised on Grew's foundations: the world-wide plant-searching travels of Linnaeus and his 'apostles', of Pallas, Humboldt, Robert Brown, Lewis and Clark, and their successors; the researches of Darwin and Mendel, and of botanists in countless laboratories; the skills of plant-breeding for the popular gardening market – these have all contributed. Today new technical equipment increases the understanding of such old problems as photosynthesis, acclimatisation, and – one of Grew's earliest interests – root-growth in darkness and compacted soil, discussed in the first chapter of his discourse on Roots.

The two parts of the Discourse on roots show Grew at his best in part 1, which records precise and detailed observations of a great variety of plants, and at his least scientific in part 2, where he put forward hypotheses, untested by experiment, about their vegetative development. The whole 'book' expands the summary account in chapter 2 of *The Anatomy of Vegetables Begun* of the extension of the radical and its 'tendency' to descend.

Part 1 of 'Roots' comprises five chapters: (1) the Origin of the root from the radicle or the stem, (2) the Skin, parenchymous and 'lignous', (3) the Bark, its 'bladders' (cells), pipes and vessels, (4) the Wood, and its sap and air vessels, (5) the Pith, and what makes up 'the whole body of a root'. Part 2 begins by justifying the study of the 'causes and ends of things' in Nature because it 'naturally leadeth us to God'. In this study Grew wrote 'I have chiefly prosecuted the Anatomical part, and I have intermixed some conjectures'. Marginal headings mark the sections of the Discourse which explain how 'the Divine Wisdom is seen in the growth of plants': first, how the ground is prepared by rain and sun, 'wet, wind and other weather', then how the sap is distributed, how the parts of the plant are nourished and formed, how they are 'situate and dispos'd', how roots are of different size and shape, differently moved and aged, how their contents and their odours, colours and tastes are made.

L.C. Miall wrote seventy-seven years ago (*The Early Naturalists*, 1912, p. 168) that Grew's 'clear and useful account of structure is sadly marred by his guesses as to function, and propensity to put forth untested speculations' instead of 'testing current interpretations by well-devised experiments', in the manner of his friend Robert Boyle.

IIc   A DISCOURSE CONCERNING MIXTURE   1675

In his first Chemistry Discourse to the Royal Society on 10 December 1674 Grew reported experiments with salts, oils, resins and gums by which he tried to discover 'the nature of bodies'. The Council on 21 January 1674/5 ordered it to be printed, and it was published by the Society's printer John Martyn in time to be listed in the Term Catalogue of 10 May 1675.

When Martyn published Grew's second chemistry discourse, *Of Luctation*, in 1678 he issued some copies combined with copies of the original issue of *Mixture*. A Latin translation was publishing in Germany in 1680 in the series of Grew's papers appearing in *Miscellanea Curiosa*.

The 'second edition' of *Mixture* was published as 'Lecture I' among the 'Several Lectures' which follow the four 'Books' in *The Anatomy of Plants* 1682; the text was only slightly revised, with a brief 'Appendix' added.

The text is divided into five Sections: in I Grew dismisses the Aristotelian doctrine of chemical mixture as 'unintelligible and unuseful', and with it the theories of Galen and the sixteenth-century chemists, Sennert alone among them 'daring to venture upon experiment'; in II he described the atomic structure of matter; in III the formation and transformation of bodies as explained by the atomic theory; and in IV the process of chemical mixture through 'congruity, weight, compression, solution, digestion, and agitation'; in V he recorded six 'instances' of chemical mixture achieved in his own experiments. He named a few contemporaries as 'curious improvers of chemical knowledge': Daniel Coxe, Robert Hooke, Robert Boyle, Thomas Willis and Walter Needham; he had himself made 'Mathew's Pill'. Richard Mathew's prescription for his diaphoretic and diuretic pill (salt of tartar dissolved in oil of turpentine) was made known in his book *The unlearned Alchymist his Antidote* 1660 (Wing M 1290), reprinted 1662 and 1663; Grew may have used the prescription for Mathew's pill from Willis's *Pharmaceutice rationalis* 1674, where this 'best corrective' is compounded of opium dissolved by salt of tartar in terebinth oil, with a little hellebore (see Keynes *Works of Sir Thomas Browne* 1963, v.4 Letters, no.77, 9 June 1679). For Daniel Coxe, missed by *DNB* and *DSB*, see Raven *John Ray naturalist* (1950) p.187 and Hunter (1982) 'F 189'.

Atomism, derived from the ancient belief in the limited divisibility of matter, was widely accepted by chemists throughout Europe during the seventeenth century. Boyle published his *Chymical experiments to illustrate the Notions of the Corpuscular Philosophy* as part 2 of *Certain Physiological Essays* in 1661, no doubt the chief inspiration of younger chemists such as Grew (Fulton 25). Newton, Grew's near contemporary, used the atomic hypothesis without a formal statement of acceptance, but published little of the chemical research which he carried out during the 1680s. (S.I. Vavilov 'Newton and the atomic theory', *Newton Tercentenary Celebrations, Royal Society 1946*, Cambridge 1947, pp.43–55).

IID THE COMPARATIVE ANATOMY OF
TRUNKS 1675

In his study of 'Trunks' Grew described the form and growth of the stems and branches of numerous plants. This was the subject of his third botanical discourse to the Royal Society, the last to be published separately. It was read at two sessions: 'The Anatomy of Trunks' on 25 February 1674/5 and 'The Vegetation of Trunks' 17 June 1675. The Society's order to print was made on 21 October, and the published book was listed in the Michaelmas Term Catalogue of 24 November 1675.

The illustrations are remarkable: nineteen large copper-plate engravings, varying from 14 × 16 cm to 17 × 23 cm mostly printed across an opening with the recto page folded; they comprise twenty-seven figures which display the differing structure of vegetable tissues in a wider range of plants than had been shown before. The first two plates contrast the 'naked eye' appearance with what could be seen through 'a good microscope', the rest are microscopic; Plate 18 fig.26 'The Lesser Common Thistle' is printed as a white image on a black ground. Ten years earlier in 1665 Hooke had shown seven examples of plant structure in his *Micrographia* (Plates 10, 11, 15 and 16).

On the title-page of 'Trunks' Grew stated prominently that his Plates were presented to the Royal Society in 1673 and 1674, no doubt to protect himself from a charge of plagiarising Malpighi whose *Anatome Plantarum*, part 1, was in course of publication when Grew's pamphlet appeared. In his dedication he reminded the President that the Society 'hath thought fit lately to give the Order for the publishing of a like undertaking by another (indeed a most accurate) hand'. A Latin translation was published in *Miscellanea Curiosa* 1680. The 'Second Edition' formed 'The Third Book' in *The Anatomy of Plants* 1682.

Called by F.J. Cole (1944) the first book in which the words 'comparative anatomy' occur in the title-page, yet Grew omitted them from his second edition. There are two dedications in the original edition of 1675: the first, to the King, was used as the general dedication of the volume in 1682, while the second, to Lord Brouncker P.R.S., remained, much abbreviated, as the dedication of 'Trunks'; the oblique reference to Malpighi's work, by now long published, was left out, but the date *August 20, 1675* was added.

The first part 'Anatomy' describes the structure of six different plant stems, from herbs to hard-woods, as seen by the 'naked' or 'bare' eye and through the microscope; in 1682 eleven stems were described. Detailed account of bark, wood and pith follows. Grew again used a homely image: 'the most proper resemblance of the whole body of a plant is to a piece of fine bone-lace, when the women are working it upon the cushion' (1675:p.37; 1682:p.121). The second part 'Vegetation' discusses movement of sap and air in the stem, the structure of their vessels and the origin of sap, movements of stems including creeping growth and 'winders' as he called spiral climbers, with the structures which underlie the suitability of various 'trunks' for human uses, from the toughness of flax to the divergent characters of the common hard-woods; also the advantage of a knowledge of structure to success in grafting.

The text was very slightly revised and enlarged for the second edition. In 1675 the chapters were analysed only in the contents-list; chapter-headings were added in 1682 and the paragraphs numbered. The plates in 1675 were unnumbered, but the figures numbered continuously and named; in 1682 the plates, increased from nineteen to twenty-three and redrawn for larger engravings, varying from 23.5 × 18 cm to 30 × 30 cm, form Plates 18–40 among the 83 Plates for the whole volume, while the figures were named but numbered only when more than one were on a single plate; the 'black' plate was not used.

IIE   EXPERIMENTS IN CONSORT OF
LUCTATION   1678

Grew made his chief contribution to chemistry in experiments, demonstrated to the Royal Society on 13 April and 1 June 1676, to test the effects of various chemical 'liquors' on many substances. More than a year later, on 15 November 1677, the Society's Council ordered the printing of these discourses. The Society's printer John Martyn, who had published *Mixture* in 1675, published the new book early in 1678; he also issued the original printings of the two pamphlets as a single book with a general title-page; this combined issue was listed in the Term Catalogue of 26 November 1677, though its imprint is 1678. A photographic facsimile of the

separate issue was published in 1962, with a valuable short introduction by Dr Michael C. Hoskin of Cambridge. The 'Second Edition'
was included in *The Anatomy of Plants* 1682 as Lecture II of 'Several
Lectures Read before the Royal Society'. A French translation was
published in Paris in 1679; it does not carry the translator's name,
and has been confused with the second edition of Louis Le Vasseur's
translation of *The Anatomy of Vegetables Begun* which came from the
same publisher in the same year, but Grew wrote in the Preface to
The Anatomy of Plants that *Luctation* was translated 'by Mons.
Mesmin a learned Physician in Paris whose version is very well
approved by those who are competent judges thereof'. This translation was included with the French version of *Anatomy of Vegetables
Begun* [etc.] published at Leiden in 1685 and again in 1691; the Paris
edition of *Luctation* was reissued in 1698.

The observations reported by Grew added little to the growing
body of chemical knowledge, and cannot be compared to the
wealth of Boyle's discoveries, but these discourses give, Dr Hoskin
wrote in the introduction to his facsimile, 'a deeper insight into the
day to day investigations, the questions asked, and the obstacles
faced' by chemists then; he noticed the range of phenomena which
Grew hoped to bring into a rational picture and his separation of
description from explanation. While seeking precision of language
neither Boyle nor Grew provided a defined terminology, as
Newton did for physics. Grew intended to produce 'a key into the
knowledge of the nature of bodies and a scrutiny of the properties of
materials used in medicine'. In this connection he analysed human
calculi, naming in his Chapter 3 (paragraph 51) 'stones often voided
by a Maid in the City of Hereford . . . sent to me by Mr Wellington
an apothecary', and added in 1682 (para. 48) 'since the first publishing . . . Mr William Matthews in Ledbury sent me part of a
stomach stone as big as a walnut of the largest size, voided by a
woman of about 82 years of age'.

Chapter I describes pouring 'menstruums' or 'liquors' on vegetable substances with their varied reactions, II on minerals, and III
on animal tissues. Martin Lister's notes in his copy at the Bodleian
Library are corroborative, unlike his earlier criticisms, except where
he wrote 'this is not true' beside Grew's statement (page 41) 'Lead-

Ore stirreth not at all with Aqua fortis'. Revision in 1682 was minimal: the Dedication was omitted and the Preface used as preliminary paragraphs before Chapter I; chapter headings and paragraph numbers were introduced. A few paragraphs were left out and others run together, with occasional changes of words or spelling.

Mesmin, Grew's French translator, was in touch with Robert Boyle for several years. Among the Boyle Letters at the Royal Society is the first sheet (four pages) of a letter in English dated from Paris on 11 [N.S.] January 1678/9 (R.S., B.L. 6, ff.39–40 attributed wrongly in the catalogue to Louis Le Vasseur, Grew's other French translator, since the conclusion and signature of the letter are missing). The Society possesses seven other autograph letters from Dr Mesmin, signed Mesmines or Mesmynes (B.L.4, ff.44–5, 46, 47–8, 50–1, 52–3, 54–5, and a letter to Theodore Haak). In the January letter the writer apologises to Boyle for delaying to write 'till I could send you a traduction I shall the next week give to the Publi[c]k of a lit[t]le English book which I confess hath been of a very delightful reading to me, and seemed to be of a very considerable usefulness. It is that lit[t]le book of Experiments in consort of the Luctation arising from the affusion of Several menstruums upon all sorts of bodies published by the worthy Dr Grew, which no doubt is the best method I can imagine by which one may enter into the knowledge of bodies. I shall not put my name to the traduction . . .'. Writing to Boyle on 24 February 1679/80, and discussing Burnet's *History of the Reformation*, he implies that he is a Protestant; in his letter of 25 May 1680 he writes in French of the pleasure he takes in cultivating his small knowledge of English and writing letters in that language; a letter of 8 December 1685 in his hand, but in the third person and unsigned, asks Boyle's help for 'le Docteur Mesmyn' and his sons to take refuge in England, undoubtedly from the persecution of the Huguenots after the Revocation of tolerance in October of that year. He may be identified with Guy Mesmin, naturalised in England on 15 April 1687 with his wife and son and later described in the Registers of the French Church of the Savoy in London as 'docteur en médecine'. (*Huguenot Society Publications* 18(1911), 189 and 26(1922), 26. If this identification is right,

Mesmin would have been in exile in England for eleven years when Barthélemy Girin re-issued Estienne Michallet's 1679 edition of his book in 1698 at Paris.

IIF   THE ANATOMY OF PLANTS   1682

Your Majesty will find, That there are Terrae Incognitae in Philosophy as well as Geography. And for so much, as lies here, it comes to pass, I know not how even in this Inquisitive Age, That I am the first, who have given a Map of the Country.

From the dedication 'To His Most Sacred Majesty Charles II, King of Great Britain, &c'.

Nehemiah Grew's great book on the development and form of plants appeared in the autumn of 1682, completing publication of twenty years' work. The volume combines revisions of his five small published books with nine unpublished 'discourses' read to the Royal Society between 1675 and 1677. *The Anatomy of Plants* and Malpighi's parallel *Anatome Plantarum*, published by the Society in two volumes 1675 and 1679, inaugurated the new science of plant physiology.

After Grew's lecture of 6 December 1676 on the form of flowers, John Wallis 'propounded that it was expedient to print all of that kind in quarto, that they might be bound together' (R.S. Journal 4). Grew preferred to collect all his lectures into one folio volume with new large illustrations. This project was delayed when the Society in 1678 commissioned him to catalogue the 'Repository'. His catalogue *Musaeum* came out in the autumn of 1681, and on 22 February 1682 the Council desired Grew 'to cause his lectures on the Anatomy of Plants to be printed together in one volume'. On 15 March they approved Proposals defining the scope and sale of the book: to include all Grew's botanical and chemical lectures, subscriptions to be paid to Grew before the end of May, and the book to appear in the Michaelmas term.

The Prospectus was printed promptly but not dated; the book itself was printed for Grew by William Rawlins and published in the autumn; it was not listed in the Term Catalogues, and was not summarised till the following summer in the *Philosophical Transactions*, v.13, no.150 (10 August 1683), 303–07.

The *Prospectus* recites the Royal Society's order to print Grew's Lectures in one volume and its proposal that the botanical papers 'be contained in about Three score and four printed Sheets and the Copper Plates will be about Three score and fifteen', the chemical lectures 'in about Thirty six printed sheets, so that the whole Volume will contain about an Hundred Sheets and about Fourscore Plates'; the completed book comprised 96 sheets and 83 plates. 'Subscribers for six English copies in Quires shall have seven', price fifteen shillings for one copy, Dutch-Demy [large paper] twenty shillings. 'Neither shall the number of copies exceed that of subscriptions given in, which it is desired may be before the end of May next, 1682. All things are so ready that the Book (God willing) shall certainly be printed in the Michaelmas term next, 1682.'

The number of copies was restricted, because *Musaeum* had been under-subscribed in the previous year. Sir Thomas Browne writing to his son Dr Edward Browne in March 1682 thanked him for 'Dr Grew's paper of Proposals', and on 29 May he wrote again 'Dr Grew sent me his proposals, which you know, on last Saturday [27 May] and I have sent him this day my own subscription, Dr Howmans, Dr Hawys & Dr Henry Bokenhams' (Browne *Works* ed. Sir G. Keynes, 4 (1964) letters 158 and 162). Grew had sent Martin Lister at York a prospectus on 8 April and thanked him on 23 May for his subscription (Bodleian Library, Lister MS 35 part 2:f.67 letter 14,f.69 letter 16).

The advance subscribers named in the Prospectus included some of the most active Fellows of the Royal Society: Brouncker, Boyle, Evelyn, Hooke, Pepys and Wren; Ashmole had subscribed, and many of Grew's physician colleagues: Edward Browne, Edward Tyson, his half-brother Henry Sampson, and Sir Thomas Millington; also several divines including William Sancroft, Archbishop of Canterbury, Thomas Tillotson a subsequent Archbishop, and John Moore afterwards Bishop of Ely. Ashmole's and Lister's copies of the published book are in the Bodleian, Pepys's in his library at Magdelene College Cambridge, John Moore's in the Royal Gift at Cambridge University Library; Evelyn's was sold in

1977 when the greater part of his library was dispersed.

Michael Hunter (1982a, pp.203 and 208 n.72) saw that Grew's publishing these 'lavish illustrated books' – this, and *Musaeum* in the previous year – by subscription, which he himself organised, put him in touch with 'a much wider circle of scientific enthusiasts than the limited group [within the Royal Society] who had supported him in 1672'. The Newcastle survey of *Book Subscription Lists* conducted by F.J.G. Robinson and P.J. Wallis shows that these were two of the earliest scientific books thus successfully financed.

The Volume comprises the Dedication to the King; the Preface, rehearsing the course of Grew's researches; *Idea* [1673], second edition, forming a general introduction; four 'Books' and seven 'Lectures'. The four Books are the second editions of (1) *The Anatomy of Plants Begun* [1672], (2) *Roots* [1673], (3) *Trunks* [1675], and (4) the unpublished Discourses on *Leaves* [26 October 1676], *Flowers* [9 November 1676], *Fruits*, and *Seeds* [both 'in the year 1677']. The seven Lectures are the second editions of (1) *Mixture* [1675] and (2) *Luctation* [1678], with the unpublished Discourses on (3) *Lixivial Salts in Plants* [March 1676], (4) *Essential and Marine Salts of Plants* [21 December 1676], (5) *Colours of Plants* [3 May 1677], (6) *Tastes chiefly in Plants* [25 March 1675] with an Appendix of the *Odours of Plants*, (7) *The Solution of Salts in Water* [18 January 1676/7].

In his dedication of 'Book 4' Grew thanked Robert Boyle for insisting that he 'should by no means omit to give examples of the mechanism of nature in all the other parts' of plants, beyond their structure. Each Discourse elaborates with new observations the summary accounts Grew had previously published.

Grew had written in 1672 'What may be the primary use of the attire [stamens] I do not now determine'; he announced his acceptance of the sexuality of plants in his discourse on Flowers (1676, but not printed till 1682), when he said 'Sir Thomas Millington told me that he conceived the Attire as the male for the generation of the seed. I immediately replied that I was of the same opinion, and gave him some reasons'. These reasons he expounded, with speculations; the important reason was his perceiving the function of pollen. He added 'that the same plant is both male and female may be believed, in that snails are such': perhaps deliberate support for Malpighi's refutation of belief in the spontaneous generation of snails maintained by Filippo Buonanni, Kircher's successor at the Jesuit College in Rome. Grew has been much praised for his 'discovery' of sex in plants; his book certainly spread the knowledge widely, but it was known to Theophrastus in antiquity and accepted by many botanical writers for at least a century before Grew. Conway Zirkle, in the introduction to his facsimile edition of *The Anatomy of Plants* (1965), recounted the sequence of publications before 1682, and clearly explained the reasons for its slow acceptance in north-west Europe.

Among the Lectures the most original are nos. 3, 4, and 7 describing rudimentary chemical analyses of plant substance. Chemistry was in its infancy and some of the terms which Grew used, even 'salt', were not closely defined till nearly a century later. Lecture 6 'Of Tastes', chapter 4 'Of the causes of taste', shows Grew's careful observations combined with hypotheses – 'supposed' explanations of 'the principles of plants' and 'conjectured' descriptions of the shapes of 'particles', referring to his former conjectures (in *Mixture*), 'and I find the learned Borelli in a book of his since published to be of the same opinion'. Giovanni Alfonso Borelli, who died at the end of 1679, had been closely in touch with Malpighi and active like him and Grew in the physiological explanation of animal structure, though also a mathematician and physicist.

Grew wrote in his Preface that he had intended 'to subjoin descriptions of imperfect plants and parasitical, marine, and sensitive plants, and the mechanism of a plant, but these things I leave to some other hand'. There are eight-three large plates, unsigned but engraved from his own drawings, which he had shown at his demonstrations. So varied a range of magnified views of plant structure displayed in dissected parts of plants had not been published before.

The Plates provide direct evidence for Grew's discoveries, showing the diversity of several 'trunks' [stems] (described at p.103ff.), the protection of buds by their scales (p.145) and of leaves by their hairs (p.247), and the connection of the pistil with the fruit (p.189).

Hooke had shown magnified drawings of plant structure in *Micrographia* (1665) on sixteen of his sixty Plates, which displayed the new world opened for observation by the microscope. He anticipated Grew in illustrating the structure of wood, leaves with their hairs and stings, and the surfaces of seeds.

A photographic reprint of the volume, slightly reduced in format, with an excellent detailed critical introduction by Dr Conway Zirkle, was published in the United States in 1965, and a microprint reproduction of the original book in 1967.

Dr Zirkle noted that Grew used the word 'anatomy' in a wide sense nearer the modern 'morphology'; Grew wished to keep the analogy between his 'anatomy of vegetables' (1672) and 'anatomy of animals' (1676). Aware that he pressed this analogy hard, he remarked in his Discourse of Flowers (chapter 5, para. 10) 'If anyone shall require the similitude to hold in every thing, he would not have a plant to resemble, but to be an animal'. Zirkle listed among Grew's contributions: his emphasis on the seed and description of germination, recognition of the bulb as a bud, absorption of water by the seed, and extension of the radicle. Grew was still collecting exotic seeds as late as 1709. Zirkle suggested also that Grew may have been attracted to botany by earlier interest as a physician in the medicinal properties of plants, for in *Musaeum* he gave much attention to such properties. But we lack evidence about his medical and botanical training between 1662 and 1671.

The Royal Society acquired 'rarities' of various kinds in its earliest years. The first dated gift in the extant catalogue was made on 18 May 1661. A year later Robert Boyle gave 'a pneumatic engine' on 22 May 1662. There were other gifts before the formal incorporation of the Society by the King in November 1662, and John Wilkins made a large gift in 1663.

A 'Repository' for these possessions was founded by Daniel Colwall, Fellow and Treasurer, when he made two gifts of £50 each, which the Council on 21 February 1666 resolved on using 'to pay for the collection of rarities formerly belonging to Mr Hubbard', usually identified as Robert Hubert who had published a catalogue of his collection in 1664. Michael Hunter (1985) records the extant manuscript, two printed issues in 1664 and a revised edition in 1665.

Gifts at the Society's meetings or sent by correspondents were regularly 'handed to Mr Hooke for the Repository'. Hooke was Curator of Experiments from 1662, and had charge of the Museum and Library as well; when Grew became Curator for the Anatomy of Plants in 1672 he seems to have shared the other curatorial duties with Hooke, and they were appointed joint Secretaries after Oldenburg died on 5 September 1677.

The Council on 18 July 1678 desired Grew to prepare a Catalogue of the Repository at his leisure; he presented his work a year later and its printing was ordered on 5 July 1679. A draft of the first six sections, describing the specimens of vertebrates and the shells, survives among Grew's papers; this draft is in Latin, but the Catalogue was published in English, though entitled *Musaeum Regalis Societatis*. The volume combines *A Catalogue & Description of the Natural and Artificial Rarities belonging to the Royal Society* with the text of Grew's lectures of 1677 on *The Comparative Anatomy of Stomachs and Guts*. He read part of his Descriptions at the meeting on 8 May 1679.

Grew used the word Musaeum in its modern connotation, but from classical times till the seventeenth century it denoted equally a study and particularly a library, the original Greek word Mouseion,

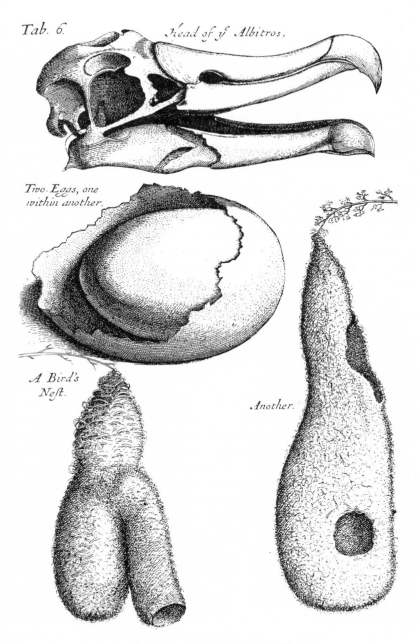

Tab. 6.   *Head of ye Albitros.*

Two Eggs, one within another.

A Bird's Nest.

Another.

PLATE I   *Musaeum Regiae Societatis* Plate 6
Skull, Eggs and Nests of Birds

29

'the Muses' shrine', having been the name of the great library at Alexandria.

Grew issued a Prospectus 'proposed at a Meeting of the Royal Society, February 16th, 1679/80' to invite subscriptions before 20 April 1680, and promising 'the impression, God willing, within or before Michaelmas Term next'. Writing to Malpighi about this time Hooke promised to send 'Dr Grew's description of our Repository soon to be published'. (Royal Society L.B.C.8,213–15 undated copy; Adelmann *Correspondence of Malpighi*, no.390 '1678 or 1680'). Publication must have been late in 1680 O.S., i.e. to March 1680/1, and the imprint is 1681. One hundred and forty-four advance subscribers are named in the Prospectus, but the printing-order evidently overestimated the demand, for the original sheets were reissued with cancel title-pages by three different publishers in 1685, 1686 and 1694.

It is now the least rare of Grew's publications; the union lists, NUC and Wing, of printed books in the chief libraries of the British Isles and North America record 105 copies of *Musaeum* and 78 of the more important *Anatomy of Plants*. In all the reissues the half-title for 'Stomachs and Guts' with its 1681 imprint was retained, but the ill-drawn portrait of Colwall, the Museum's founder, was replaced by Robert White's excellent engraving; this also is dated 1681, though it was not used till 1685 (*see* Plate 7). In his Epistle Dedicatory to Colwall 'your Voluntary Undertaking for the Engraving of the Plates' was gratefully acknowledged by Grew.

The Prospectus, issued early in 1681, promised 'a perfect Catalogue of all the Rarities belonging to the Royal Society at Gresham Colledge, containing Descriptions . . . comprized in about Fourscore Printed Sheets; with the Figures in about Twenty five half Sheets from Copper Plates . . . also a treatise of the Stomachs and Guts of about Thirty sorts of Animals in about Ten Printed Sheets with the Figures in about Five half Sheets, now to be published by the Order of the Society . . . Subscribers for six copies will receive seven, and the price will amount to about Ten shillings. Subscriptions are to be notified to Doctor Grew at or before the Twentieth of April next. The Copies shall not exceed the number of Subscriptions. . . . payment to Doctor Grew at his House near Warwick-Court in

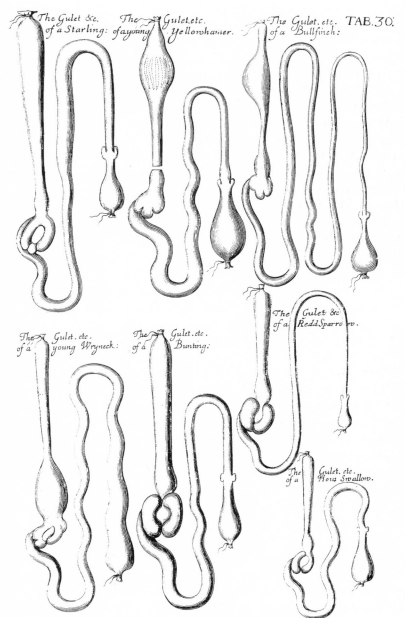

The Gulet &c.
of a Starling:

The Gulet. etc.
of a young Yellowhamer.

The Gulet. etc.
of a Bullfinch:

TAB. 30.

The Gulet. etc.
of a young Wryneck:

The Gulet. etc.
of a Bunting:

The Gulet &c
of a Redd Sparrow.

The Gulet. etc.
of a Hous Swallow.

PLATE 2   *Comparative Anatomy of Stomachs and Guts*
Musaeum Plate 30 Gullets of Birds

31

Warwick-Lane, London. Although the Book be prized no higher than at about Ten Shillings yet at the usual and ordinary price of Books of this nature it will be worth about Eighteen Shillings'.

The list of advance subscribers includes twenty-seven of the eighty-three donors named in the published book, several leading Fellows of the Royal Society – Boyle, Hooke, Petty and Wren, great officers of Church and State – Sancroft, Archbishop of Canterbury, Compton, Bishop of London, and Ward, Bishop of Salisbury, the Lord Chancellor, the Lord Privy Seal, the Lord Mayor, and many physicians – Edward Browne, Croone, Ent, Goodall, Merret, Millington, Needham, Sampson (Grew's half-brother), Scarburgh, Sydenham and Tyson, with others less well remembered. (See Appendix I).

IIIA   THE ROYAL SOCIETY'S MUSEUM AND ITS EXHIBITS

The 'Repository' reflected the policy of Henry Oldenburg and the genius of Robert Hooke, the Society's permanent officials. Oldenburg's international correspondence made the Society a clearing-house of new knowledge, while Hooke arranged regular demonstrations of discoveries and new instruments. Scientific and practical devices which had been exhibited at the meetings were usually handed to Hooke for the Repository; they formed its most original and important part. Early in this century Martha Ornstein (1875–1915) in her pioneer study *The Role of Scientific Societies in the Seventeenth Century* (1913, reprinted 1928 and 1938) asserted correctly that 'the rarities were comparatively on a lower level than the instruments'. These instruments were kept by the Royal Society when most of the Repository was transferred to the British Museum in 1781, exactly a century after Grew published his catalogue *Musaeum Regalis Societatis*, which included descriptions 'Of Instruments relating to Natural Philosophy' (pp.357–68). Robert Boyle, Christopher Wren and Robert Hooke were foremost among the Fellows who described new instruments in the *Philosophical Transactions* or deposited them in the Repository. These included Boyle's air-pumps and condensing-engines, Wren's rain-gauge, and many of Hooke's contrivances. The rain-gauge was also used in the weather-clock devised by Wren and Hooke which 'advantageously' combined clock, barometer, thermometer, weather-cock, rain-gauge and hygroscope, each registering on 'a paper falling off of a roller' worked by the clock. Hooke's list of his own inventions survives (R.S.,C.P.XX, nos.1–97, printed by Geoffrey Keynes *Bibliography of Hooke* 1969 pp.76–84); he placed in the Repository his lamp-furnace for hatching eggs to observe the process of generation, with its semi-cylindrical lamp 'for poising the liquor which is to keep the furnace at the same height'; his model of an iron instrument to fetch earth or other bodies from the depth of the sea; a model of an eye in which the humours are represented by glasses; and a burning-glass of his own devising. John Wallis gave a hollow burning-glass and several 'otacoustics' (ear-trumpets).

There were also a large microscope with three 'glasses' and a lesser one 'somewhat more manageable'; Grew recommended making use of both and praised the very small lenses ground by John Malling, presumably to resemble Leeuwenhoeck's. Under the subsidiary heading 'Of Things relating to the Mathematics' the list continues with Isaac Newton's reflecting telescope, the subject of important articles in the *Notes and Records of the Royal Society* (A.A. Mills & P.J. Turvey 33 (1978), 133–56; A.D.C. Simpson 38 (1984), 187–214). Here too were Hooke's 'Instrument for working questions by multiplying and dividing', John Wilkins's 'waywiser' – a surveyor's wheel or distance-meter as used for Grew's economic project of 1711. Wilkins also gave a model of John Wallis's 'geometrical flat floor'. Wren deposited an 'instrument to demonstrate how far against the tide a ship may sail', and his 'terella' or spherical magnet; 'loadstones' were listed earlier in the catalogue (pages 317–18) with an account of Theodore Haak's experiments in recharging them.

Later in *Musaeum* (pp.353–56) Grew recorded a preparation of phosphorus 'made by Dr Slare and by him given to the Royal Society, April 3, 1679'. Frederick Slare M.D., F.R.S. was a protégé of Robert Boyle and had confected this for his 'great and noble patron' who examined it with Prince Rupert. Slare's description, which Grew printed, quoted Haak's observation of the varying light in the phosphorus which showed 'a fiery complexion as the sun approaches the meridian, while expos'd to a candle a pale luminous colour, but in clear moonshine not luminous in the least'. 'As to his [Slare's] conjectures' Grew commented

> as to the subject and cause of light in this phosphorus, because he hath desired my opinion I subjoyn it: What it is that gives the light it seems hard to say – the cretaceous salt, the nitrous salt, or some igneous particles incorporated with them? How the particles give light? It should first be stated what light is: whether it be a body? all one as to ask, whether a quality be a body? The question ought to be whether there be any body which is the peculiar subject of light? Whether it be any other adjunct besides motion? Whether there is one peculiar motion at least for a musical sound, so another for light? ... Some experiments seem to favour each of these questions.

Grew was an instinctive discoverer who knew what he was looking for, but contemporary chemistry and physics could not furnish him with the knowledge necessary to prove his hypotheses.

Under the heading 'Warfare' Grew entered Lord Brouncker's recoiling gun, Prince Rupert's assayer for gunpowder, Wilkins's wind-gun, and Dudley Palmer's seven-shot gun. There were also exotic weapons from North and South America and the East and West Indies, with a canoe from Greenland. More homely inventions were Charles Howard's saffron kiln, Hooke's cider-press, and Wren's box-hive.

For the rest the Repository comprised a remarkably large assembly of animal, plant, mineral and fossil specimens. There was a smaller group of exotic coins and a few antiquities. These also came under the heading 'Of Things relating to the Mathematicks and some Mechanicks', including further exotic artefacts such as 'A West-Indian bow, arrows and quiver', 'A Siam drum ... the parchment made of a fishes skin', [Amer-]Indian birch-bark utensils, 'An Indian Peruque, made not of hair but feathers, a mantle also of feathers' (*Musaeum*, p.367).

John Winthrop F.R.S., Governor of Connecticut, had sent the Society samples of 'mayz or Indian wheat'; Grew reported that the Indians eat it mixed with grains of chocolate. His continuing interest in ethnography is attested by the queries he sent a decade later to friends in Massachusetts about the Indians there (see Chapter VIII).

Grew's classification of the animal specimens was uncertain. He placed spiders among insects, and listed some caterpillars as distinct species while correctly describing the cycle 'worm, aurelia, butterfly' for the silk-moth (*Bombyx*). He wrote 'the bat stands in the rear of beasts and in the front of birds', but described bats under the heading of 'Land fowls'. Similarly he placed Seals and Porpoises among 'Viviparous Fishes', though the dissections he recorded show that he knew they are mammals.

Beside Robert Hubert's exotic animal exhibits Grew found in the Repository, as Michael Hunter noticed (1985, pp.153–66), several beaver and tiger specimens. He was ill-informed or unaware that the name 'tiger' was given to various species when he wrote that 'Tigers abound in Mexico, Brazil, and the East Indies'; and he belied his normal good sense in repeating the 'Physiologus' legend that 'the Leopard is begotten by a Lion upon a Panther'.

Beaver were still frequent in north-west Europe, breeding as Grew wrote 'in Italy, France and other places, but our best Castor is from Russia'; this aromatic secretion *Castoreum* was used medicinally according to him 'in hysterical and comatose cases'. By the nineteenth century the animal was restricted to Canada and north Asiatic Russia. The beaver's anatomy was known to the Graeco-Roman naturalists and described by several Renaissance anatomists – Grew named 'Aldrovandi, Worm and others' and mentioned 'the description in Gesner'.

The best reports of beaver dissections in his own time were those by Claude Perrault in Paris (published 1669), which Grew quoted from the summary in the *Philosophical Transactions* 49 (1669), 991–96, and by J. J. Wepfer at Schaffhausen on the upper Rhine (1667, published 1671). Wepfer had dissected two male specimens, the first with help from Grew's half-brother Henry Sampson. The peculiar features were duly noticed: the strong incisor teeth, the tail, the os penis, and the 'castor' sacs, previously confused with the testes. (Cole *History* pp.71 and 365, *Catalogue* nos 671 and 778). Grew described specimens in the Repository of the incisors and the tail, with 'the pistlebone, so I find it inscribed', but merely mentions the 'castor-bags'.

His personal interests as physician and botanist appear in his discussion of the uses of the plants he catalogued, and at the end of the volume (p.387) he provided a special index of medicines arranged by their uses. At pp.227 and 376 he described a specimen and drawing of the Chinese Ginseng root given to the Society in 1679 by Dr Andrew Clench; John Appleby has shown that this exotic medicament interested many Fellows of the Society from 1666 to 1788 (*Notes and Records of the R.S.* 37 (1983), 121–45).

Among exhibits illustrating chemistry Grew mentioned the 'Fixed Salts' of various plants 'by Me prepared and given, together with a Discourse before the Royal Society which I propose to publish'. He had given three 'Lectures' on salts in plants in 1676 and 1677 which he included in *The Anatomy of Plants* (1682).

Many of the exhibits have not survived the passage of the years and the various removals of the collection. The first entry in *Musaeum* describes the Egyptian mummy given by the Duke of Norfolk. Though in bad condition, 'converted into a black rosin', when Grew saw it, it was examined by a group of anatomists in 1774, but seems not to have been sent to the British Museum in 1781, where the earliest gift of a mummy came in 1837 according to the *Catalogue of Mummies in the British Museum* by W.R. Dawson and P.H.K. Gray (1968). The use of 'mummy' – oils or resins from mummy-wrappings – as medicine had been practised for many years, but Grew dismissed it: 'let them see to it, that dare trust to old gums which have long since lost their virtue'.

There were a few other exhibits of human tissue, including a piece of tanned skin, about which Grew wrote 'Bartholine (Historiar, cent. 3) shews the way of tanning a humane skin, I believe it may be tann'd by all the ways which are us'd upon other skins'. He also discussed, with inconclusive teleological explanations, the comparative measurement of male and female human skeletons. (*Musaeum* pp.4–7).

In the collection of fossils he described the 'glossopetrae' correctly as sharks' teeth, without mentioning that this was a comparatively recent identification of the so-called dragons' tongue stones published by Nicolaus Steno in 1669 and translated into English by Henry Oldenburg in 1671. Grew recognised the fossil 'hippopotamus' tooth discovered in Kent by William Somers as a rhinoceros tooth (pp.14–15, 29–30 and Plate 19). This *Rhinoceros antiquitatis* specimen is still preserved in the Palaeontology department of the British Museum (Natural History), with the 'petrifyed tooth of a sea animal' (shown also in *Musaeum* Plate 19), subsequently identified as a molar of the mastodon *Tetrabelodon angustidens* (W.N. Edwards *The History of Palaeontology*, B.M.(N.H.) 1976 pp.50–51). He knew that fossils are relics of earlier times, but seems not to have envisaged the long time-sequence of successive deposits which Hooke had realised.

Musaeum is a record of other men's inventions and collecting, whereas *The Anatomy of Plants* describes Grew's own discoveries and consequently was much more influential. *Musaeum* is well-known as a mirror of the interests of early Fellows of the Royal Society, but Grew's scheme of arrangement, manner of description, and frequent expression of opinion have been little considered.

Michael Hunter (1985, p.165) noticed that Grew 'urged valuation of the ordinary' in preference to the unusual rarities of the virtuosi, instancing his commonsense explanation of the apparent miracle of branches growing in the shape of crosses that they must have been 'bound together when young, and so by a kind of ingrafting became coalescent' (*Musaeum*, p.184), or his dismissal of the superstition of the 'Ship-halter' fish, Echineis or Remora (pp.104–06): 'I meet with no tolerable description anywhere . . . though the moon be made of a green cheese, yet it is not the only nest of maggots'.

His preface explains the decisions he took for classifying, naming and describing, but the text shows a few surprising errors of judgement. He has been criticised especially for his ignorance about shells, of which there was a large display. He was at a loss to identify many of the evidently unlabelled eggs which he merely describes, and suggested that an egg 'somewhat bigger than that of a turkey is as I take it the egg of a Guillemot' (*Musaeum*, pp.78–79). However he deserves praise for diligence and speed in compiling his catalogue, and for acuity in identifying undescribed or inaccurately labelled specimens. He wrote in his Preface: 'In the descriptions I have taken care to rectifie the mistakes . . . given us by other hands. Not to transcribe any, . . . but having noted something especial, to refer to the author. Where there is no description, or too short, or the faults many, to give one at large. For the doing of all which, what the trouble of comparing books together hath been, Post Deum Immortalem Ipse novi'.

Throughout *Musaeum* there are references to previous descriptions of comparable specimens. The predecessors on whom Grew chiefly relied were named in his Prospectus: the Museum Catalogues of Francesco Calceolari, Ferdinando Cospi, the Jesuit College at Rome [Athanasius Kircher], Lodovico Settala at Milan, and Ole Worm at Copenhagen, with the writings of Ulisse Aldrovandi, Conrad Gesner, Guillaume Rondelet, Ippolito Salviani 'and other authors'. Presumably Grew used books in the library which Henry Howard, afterwards 6th Duke of Norfolk, gave to the Society in 1667; William Perry was cataloguing this library while Grew was cataloguing the Repository, but his *Bibliotheca Norfolciana* remains unpublished in the Royal Society's archives. The greater part of the library was transferred in 1830–35 to the British Museum and the remnant of the printed books sold at Sotheby's in 1925. The library had been formed by the Duke's grandfather Lord Arundel; its history is outlined in F.C. Springell's book *Connoisseur Diplomat* (1963) p.110. Dr Michael Hunter (1985) found from the Royal Society account-books that the Society bought books specially for Grew's use during 1679–81.

The Society moved from Gresham College to Crane Court on the north side of Fleet Street, next before Fetter Lane, in 1711 a few months before Grew's death at Racket Court a quarter mile east. As planned by Wren five years earlier, the 'Books, Rarities and Instruments' were transferred to the 'new Repository' there. (J.A. Bennett 'Wren's last building?' *Notes & Record of the Royal Society* 27 (1972),' 107–15.)

The Royal Society added continuously to its museum till about 1765. The later history has been well summarised, with particular reference to the scientific instruments, by A.D.C. Simpson: 'Newton's telescope and the cataloguing of the Royal Society's Repository', *Notes and Records of the R.S.* 38 (1984), 187–214. Hans Sloane during his long Presidency (1727–53) kept for his own collection many specimens which would have entered the Repository in a more public-spirited age than that lost half-century of English science; but Sloane made amends by leaving his manifold treasures to found the British Museum.

When the Royal Society moved from Crane Court to Somerset House in 1781 the natural history collection and other rarities were transferred to the British Museum, as noted above; some specimens had been transferred in 1772 and 1774 (A.E. Gunther 'The Royal Society and the foundation of the British Museum 1753–1781' *Notes and Records of the R.S.* 33 (1979), 207–16). The British Museum sold part of the Royal Society's collection in 1809 for £175 to the Royal College of Surgeons for the Hunterian Museum, which the College holds in Trust for the Nation. The Conservator William Clift (F.R.S. 1823) listed fifty-four groups of specimens transferred (R.C.S., MS 276.h.a. 29) and wrote some Memoranda of the transfer (B.M.N.H. General Library, MS SB 9 C) in which he described finding 'in the souterraine of Montague House [the B.M.] an immense congregation of bottles apparently accumulated from the days of Dr Hooke ... monsters, mice foetuses without number, common English snakes by the score ... Most of these were thrown away when they came to the College'. The labels were mostly illegible, 'many of the specimens were certainly Sloane's ... Some had been part of the Rarities described by Nehemiah Grew, vizt: Four Tables showing the Arterial, Venous and Nervous Systems prepared at Padua, presented by John Evelyn (*Musaeum* p.4); a wreathed elephant's tusk (*Musaeum* p.31) given by Thomas Crispe from the Royal African Company; the skeleton of a very large crocodile (*Musaeum* p.42 and Plate 4) given by Sir Robert Southwell'. The 'Evelyn Tables', still exhibited in the Hunterian Museum, are human dissection specimens mounted on four panels. Evelyn in his *Diary* described their purchase at Padua, February 1646, and his gift of them to the Royal Society 27 October 1667; his 'Letter to William Cowper relating to the Anatomical Tables' was printed in *Philosophical Transactions* 23 (1702), 1177–79. Sir Arthur Keith explained their arrangement in his *Illustrated Guide to the Museum of the R.C.S. of England* (1910, pp.21–22), and Sir Geoffrey Keynes recorded their history in *John Evelyn, a Bibliography*, 2nd ed. (1968, pp.260 and 285); the history of the similar Tabulae Harveianae at the Royal College of Physicians was told in Archibald Malloch's *Finch and Baines* (1917).

The Repository specimens retained at the British Museum were moved to the new Natural History Division at South Kensington in 1881, a second century since *Musaeum* was published. A few of the original 'rarities', besides the fossils already mentioned, may still be seen there. (B.M.N.H.: *History of the Collections* 2 (1902), p.4 'Zoology' by Oldfield Thomas).

IIIb   MUSAEUM 1681——COMPARATIVE ANATOMY OF STOMACHS AND GUTS

Grew 'subjoined' to *Musaeum* the lectures on the comparative anatomy of stomachs and guts which he had read to the Royal Society four years earlier. He explained in his opening paragraph that he intended to 'omit most of what is already noted by Anatomists, and principally speak of those things which have hitherto been unobserved'. He gave, in his first discourse, a summary of the work of earlier comparative anatomists, which he did not print, restricting himself in 1681 to reporting his own observations. He probably hoped to continue his research through the separate 'systems' of the animal body, as John Hunter did a century later, but completed only this study of 'stomachs and guts'. F.J. Cole gave a detailed analysis of the discoveries and shortcomings of these lectures in his *History of Comparative Anatomy* (1944) chapter 22 on Grew, with further discussion of *Musaeum* in chapters 33 and 35.

Grew certainly produced new knowledge, and was even more successful in showing the true value of comparison. Earlier anatomists examined animal structure for the light it threw on human anatomy, seeing analogies rather than comparisons.

When he described the animal specimens in the Repository (*Musaeum* pp.10–48) Grew divided mammals and reptiles, except 'Serpents', between 'viviparous' and 'oviparous quardrupeds'. In these lectures of 1677 on Stomachs and Guts he compared the variations of structure in this system from a large number of different animals, both herbivores and carnivores, classified by their feeding habits, deducing from his own observations that the physiological function of an organ must be understood to explain morphological difference in superficially similar animals. Yet, as F.J. Cole wrote (*History* 1944 pp.246–47) this classification, correct in principle, does not 'harmonise with the facts Grew himself discovered – identity of structure of the digestive system organs in mole and hedgehog, which he had observed and described, is meaningless if these animals are in different groups, and diversity is equally ignored if the hedgehog is made to lie with the rodents'.

He seems to have been the first anatomist to explain the mechanism of digestion in the specialised organs of ruminants, from his dissection of a sheep; this compound stomach had been described by Marcantonio Severino in his *Zootomia* (1645), but Grew 'first fully recognised the implications of comparison with the simple stomach of other animals including man' (Cole *op.cit.* pp.14–15 and 146–47). Complete determination of the difference of carnivorous and herbivorous digestion was made one hundred and twenty years later by Claude Bernard during his experiments to explain the role of gastric and pancreatic juices in the digestive process (1846–49).

Grew described thirty-eight animals and mentioned thirty more, but illustrated only twenty-eight (eleven mammals, fourteen birds and three fish). Almost all were common English species including domestic mammals and birds. 'A squirrel that I opened was a Virginian, smaller than the European', evidently the North American red squirrel *Tamiasciurus hudsoniana*, which is three inches shorter. Among birds he noticed similarities and differences in related species; he had dissected a cassowary, but did not differentiate cassowary from emu, as was usual before the exploration of New South Wales late in the eighteenth century; both names occur in English from the early 1600s. Living birds of one or other species were brought to England from 1596 onwards; both William Harvey and John Ray had seen them; John Tradescant the younger and Sir Thomas Browne each had an egg of cassowary or emu. Grew used the generic name Fitchet for the Stoat to which also he compared the Ferret, though both 'Fitchet' and Ferret are usually connected with the Polecat.

In his account of the Cat he wrote 'To its guts I suppose those of a Leopard, Tiger and Lion may have some analogy' and recorded that he had not dissected a Civet-cat. The Dog he compared to 'all ossivorous quadrupeds', while with the Rat he compared the Mouse and contrasted the Shrew-mouse; he also compared the Pig's colon with those of the Horse, Ass, and Coney – as he sometimes called the Rabbit (his usual term). With the Calf he mentions Ox and Cows.

The words 'carnivore' and insectivore' were already used before Grew's time but he was probably the first anatomist to add the parallel 'frugivorous' 'graminivorous' and 'ossivorous'. He used no Latin nomenclature except in adding the synonym 'Avicula Anadavadensis' at his mention of the Twite, apparently confusing a familiar European bird with the Indian song-bird, called from Amadabad the 'amadavat', which became a popular cage-bird among English ladies a century after Grew's time. For the shape of certain guts he used the adjectives 'helic' and 'conic' rather than the current 'helical' and 'conical'.

Grew suggested (p.41) that the red colour of muscle in certain animals where the flesh is white must be due to their 'strong and continual motion'. In describing the few specimens of human viscera in *Musaeum* (pp.7–8) he named the physicians – Jan Swammerdam, Edmund King, Regnier de Graaf and others – who had prepared or presented them, and quoted William Cole's observation, published in the *Philosophical Transactions* no.125, that 'the valvulae conniventes in the jejunum and the muscular inner membrane of the guts are continu'd in a spiral line all along from the stomach to the very anus'. When, however, he came to write here of the glands he did not quote Thomas Wharton's comprehensive *Adenographia* of 1656; Wharton, who had died in 1673 before Grew's lectures were delivered, was one of the old group of College of Physicians researchers who resented the rise of the Royal Society.

Grew used homely similes in his plant studies, and here too he wrote, for instance, that the rabbit's caecum 'being blown up looks like those skins of iceing glass [isenglas] formerly used for transparent flower-works'. Discussing somewhat inconclusively the glands in the intestine and 'the mucus which they spew' he suggested that 'the glands are a great means to prevent feavers'.

Lecturing early in 1677 he anticipated discoveries published later that year, though probably made long before, by Francis Glisson (*De ventriculis et intestinis*). He refuted an opinion of 'the learned Dr

Glisson' who died in this year aged eighty, while in 1672 he had gratefully recalled Glisson's influence on his early work. When he published these lectures in 1681 he printed at the beginning 'An Advertisement to the Reader' claiming priority over Johann Conrad Peyer of Schaffhausen in recognising the lymphoid follicles, sometimes called 'Peyer's patches', on the mucosa of the small intestine. Peyer had published his observation of these 'Noduli lymphatici aggregati' in his *De Glandulis Intestinorum* (1677) and repeated his account in his *Parerga anatomica* in this year 1681. Like Grew he also published a study of ruminants: *Merycologia* (1685); Grew's half-brother Henry Sampson had studied anatomy under Peyer's master J. J. Wepfer at Schaffhausen in 1667 (their beaver dissection is mentioned above). Grew also quoted, with slight corrections, observations by Thomas Willis and Walter Needham and 'a slip of Francis Willughby's most accurate pen' (pp.17, 21, 22, 26, 32, 35).

Grew gave no measurements in the text, merely imprecise comparisons with seeds or coins, but provided an inch-scale on his illustrations. 'Some Notes upon the Tables' (p.43) indicate that in Plates 23, 24, 26 and 28 some of 'the Guts are inverted to shew their glands'.

The Royal Society possesses a manuscript report of Grew's first lecture, which varies little from the printed text; indeed he retained in print a few phrases from the lecture-room, such as 'dissections presented before this Honourable Presence'.

New Experiments concerning Sea-water made fresh was published without Grew's name in 1683. He wrote it to support Robert Fitzgerald's new method of distilling sea-water to make it drinkable, which Fitzgerald was promoting by his own pamphlet *Salt-water sweetned*; in 1675 a Patent for a similar method had been granted to William Walcot.

Grew's pamphlet is ostensibly an address to the King, expounding the value to the health of His Majesty's subjects of the Patent granted to Robert Fitzgerald for his new method. Fitzgerald had first asked for medical support from Richard Lower who sent him to Grew; in thanking Grew later, he told him that 'His Majesty received and read with great satisfaction your ingenious Treatise of the Sea Water'. Grew had demonstrated salts from various London waters to the Royal Society in 1679; he included in this pamphlet part of his unpublished 'Discourses', comparing water from Chancery Lane and Christ's Hospital pumps and Cheapside conduit with Thames and New River water, and added new analyses of distilled spring water and of sea-water both fresh and distilled by the new 'engine', and of course referred to Fitzgerald's 'late Book'. He recommended the use of this distilled water, particularly as a preventive of the scurvies and dropsies caused in long voyages by bad water as well as bad meat. Grew signed the Approbation from the College of Physicians which is included in Fitzgerald's pamphlet, and in his own pamphlet stated 'I have no share in the Profit or in the Credit of the Experiment'.

Stephen Hales, 'engaging in the attempt to make distilled sea-water wholesome' sixty years later, gave a summary in the preface to his *Philosophical Experiments* (1739) of 'what was done by Mr Walcot and Mr Fitzgerald in King Charles the Second's time', but did not mention Grew's contribution. Hales was sceptical about the secret ingredients which his predecessors claimed to have used, and concluded that 'the principal thing was only Distillation'. Fitzgerald was a nephew of the great Robert Boyle, and Hales recalled that he had acted 'upon Mr Boyle's encouragement'. Knowing from his own experiments that distillation gave variable results, Hales added

'Tis probable that Mr Boyle might happen to try with a solution of silver some good Water of Mr Fitzgerald's preparing, who might bring him the best he had. For it is not to be supposed that so worthy and good a man, as Mr Boyle was, would impose a falsehood on the world for the sake of anyone whatsoever'. Boyle had published *Tracts concerning Observations about the Saltness of the Sea* in 1674, in which the 'Advertisement' tells that he became interested in the welfare of sailors and the provision of fresh water at sea from his service in the Court of the East India Company and the Council for the English Colonies in America. Fitzgerald's *Salt-water sweetned* includes 'A Letter of Mr Boyle's to the learned Mr John Beale'. Boyle's 'Letter' recounted six characteristics from his 'Trials' of the distilled water, which convinced him that Fitzgerald's claim of its 'potableness and wholesomeness' was justified; Boyle had also satisfied the King's 'judicious objection against practicableness' in its production. In the *Philosophical Transactions* of 30 October 1683 Boyle published an 'Appendix to his Letter', describing his 'Way of examining Water as to freshness and saltness'.

Grew dedicated his pamphlet to the King, with assurance that the new method was easy and safe in the hands of 'the Gentlemen who have your Majesty's Patent' and had been approved by Boyle and the Physicians. He described his own analysis of various waters and indicated the value of the new method for sailors during long voyages or when becalmed, especially in the tropics.

Fitzgerald's pamphlet answered nine questions about his new method: (1) one seaman could attend several 'instruments' and produce adequate quantities of water for a ship each day, where (2) for distilling by fire 'a skilful chymist' would be required, (3) the instrument is never likely to go out of order, unlike the ordinary 'engine', (4) it will work in the worst weather, (5) there is no danger of fire or smoke, (6) it is cheap, may last many years, and can be used on shore, (7) the old method entailed storing expensive coal on board, taking up ten times as much space as one day's water, (8) the new 'ingredients' are cheaper and take much less space than the old, (9) Dr King's analysis shows that this 'dis-salted' sea-water is not 'unwholesom'. He included favourable testimony from eight peers and twenty-two Fellows of the College of Physicians, but the 'instrument' and 'ingredients' were not described. Fitzgerald's letter thanking Grew for his help was printed by Josiah Peter in his vindication of Grew during the Epsom Salt controversy *Truth in Opposition to Falshood* 1701, p.17.

Robert Fitzgerald was the younger son of the 17th Earl of Kildare, premier peer of Ireland, and his Countess, Lady Jane Boyle daughter of the first Earl of Cork, the most prominent Elizabethan settler; hence he had direct friendship with his uncle Robert Boyle 'the father of chemistry' and ready access to the King. William Walcot had received the Patent for his method of 'purifying corrupt water and making sea-water fresh, clear and wholesom' on 28 October 1675 (Woodcroft *Index of Patentees*, no.184). Walcot was aggrieved by the issue of a new Patent on 9 June 1683 to Fitzgerald and his associates Theophilus Oglethorpe, William Bridgman, Thomas Maule and Patrick Trant for their 'Engines for purifying salt and brackish water, making it sweet and fit for drinking and purposes of cooking' (Woodcroft, no.226).

The controversy between Walcot and Fitzgerald before a Committee of the House of Lords is recorded in the Historical Manuscripts Commission Report which calendars the *Manuscripts of the House of Lords*.

'A Bill for making Salt-water fresh' was introduced in the Lords on 13 March 1677/8 (as Mr D.J. Johnson, Deputy Clerk of the Records in the House of Lords has kindly told me) but there is now no information about it; this was intended, I assume, to lead to Walcot's patent receiving the authority of an Act of Parliament.

On 28 February 1692/3 a Committee of the Lords considered 'A Bill to restore to Mr Walcot his rights under the Letters Patent of 1675 interfered with by other Letters Patent in 1683'; at a second meeting on 7 March leave was given to Robert Fitzgerald on his Petition (HL,MS 729) to be heard against this Bill. On 9 March his counsel Sir Bartholomew Shower 'says that Mr Walcot's Patent is expired. He desires no right to be confirmed, but a right to be given him ... We [Fitzgerald's partnership] pray a saving of our right to our invention', and Mr Holles produced an Order of Council dated

31 October 1683 for vacating Mr Walcot's Patent. Sir Thomas Powys, for Mr Walcot, said 'the invention must be lost unless we have an Act'. On 10 March Fitzgerald 'offered a Proposal... it was read and given back to him... The Bill was negatived' (HL,MS 723).

A year later on 21 March 1693/4 Walcot's Bill was considered again, and after a further year the Committee received on 21 February 1694/5 an amended draft of a Salt Water Act; on 13 March Walcot attended with Counsel; letters from Fitzgerald were read: 'he would not trouble the Committee, but if his four years in his Patent might be saved, it would be a kindness' (HL,MS 910). The quarrel, however, was continued in pamphlets till 1702.

Three editions of Fitzgerald's and Grew's pamphlets were issued in 1683, also a Dublin printing of Fitzgerald's, and during the next year and a half they were reissued to their tenth editions, more or less simultaneously and sometimes combined, with complete or partial translations into Latin, French, German and Spanish.

Grew's pamphlet was issued to its fifth edition as 'By a Fellow of the Colledge of Physicians and of the Royal Society'; the sixth and seventh editions are not now known, but his name appears in the eighth dated 12 August 1684.

There were also supporting pamphlets and broadsheets: The most important of these were:

(1) 'The Agreements and Conditions' to be contracted between the Patentees and 'everyone buying their Engines and Ingredients' with the charges to be paid at their office in Lothbury near the Royal Exchange; 'all parties may have delivered to them two Treatises wrote on the subject in most languages at the Office... or at The Ship in Cornhill'; this was issued as a broadsheet, but subsequently enlarged and included with the 8th, 9th and 10th editions of Grew's *New Experiments*.

(2) The 'Certificates of certain Captains', to which was added in the 10th edition a third Certificate. The ships' Captains who wrote evidence of their successful use of Fitzgerald's distilled water were John Kempthorn (1684) and Robert Crauford (1685), with Randal Macdonnel (1683) added later.

(3) 'A Brief of Two Treatises', i.e. a broadsheet summary of Grew's and Fitzgerald's pamphlets, of which two issues and an Italian translation are known.

V   BITTER PURGING SALT   1695

This analysis of Epsom Water renews research which Grew had begun 'more than fifteen years past', for in June and July 1679 he had read papers to the Royal Society on the mineral and other waters 'in and about London' and demonstrated the salts he had extracted (Royal Society Journal 5 June and 3 July 1678 papers read; John Evelyn *Diary* 19 June 1679 demonstration of 'salts extracted and produced before us and examined'). Josiah Peter *Truth against Falshood* 1701 quoted testimonies from Wren and Sloane that Grew gave these demonstrations, and from Hooke that the papers were not detailed in the Journal because 'we looked upon the experiment as new' and Grew intended to revise the papers.

Epsom waters were discovered about 1618 (A. Sakula 1982). Till the mid-century physicians sent their patients to drink the waters at Epsom. Dorothy Osborne went there in 1652 'to be rid of a scurvy spleen', and wrote in the following summer to William Temple 'Did you drink directly from the well?... At this time of year the well is so low that you can hardly get any but what is thick and troubled. When it had stood all night the bottom of the vessel would be covered an inch thick with a white clay'. (Dorothy Osborne *The Letters to Sir William Temple*, ed. G.C. Moore Smith 1928, nos. 3, 26, 31 & 32).

'About thirty years since', Grew wrote here in 1695, 'many of the inhabitants of London had Epsom water sent to them, and other waters from near London grew into request and use'. He was among the first to analyse local mineral waters, and had used a part of his earlier papers in *Sea-Water made fresh* (1683); he now used a further part, listing salts from mineral springs within about fifteen miles from London: Barnet, Northaw, Acton, Cobham (Surrey), Dulwich, and Streatham, besides Epsom.

During the 1680s several other Fellows of the Royal Society, among them Boyle, Petty and Aubrey, made further analyses, but Grew's *Treatise* is the most complete chemical and medical discussion of a particular water.

Chemical terminology being as yet informal, the word 'salt' was used for any solid soluble non-inflammable substance without precise definition. Grew described the salt of Epsom water as 'lixivial alkaline'; it was subsequently identified as Magnesium sulphate, and he invented a method of preparing the salt for medicinal use.

The Epsom Water book is Grew's one extensive publication in Latin, the conventional language for a medical subject. It was approved by the Royal Society on 27 March and by the Royal College of Physicians on 3 May 1695, and published before June by Smith and Walford at the Prince's Arms in St Paul's Churchyard.

Two years later in the spring of 1697 Francis Moult, an apothecary with a shop in Watling Street, published an English translation 'which I intend', he wrote 'to give to those who buy any quantities of the Salt', assuming Grew's 'pardon for my translating it without his knowledge'. When Grew failed to obtain apology from Moult he authorised Dr Joseph Bridges to publish a new translation in the autumn with a narrative by Josiah Peter, dated 4 August, describing an abortive negotiation with Moult on Grew's behalf, and 'animadversions on a late corrupt translation'. Grew had hoped that Moult would apologise and withdraw, but when Peter brought him to Grew, who remonstrated with him, 'instead of giving a modest and sensible answer, he pertly said "And, Doctor, what is the Consequence of all this?" and thus discovered himself a Person of great Impudence and Ignorance'. The controversy was a private extension of the current dispute between the College of Physicians and the Society of Apothecaries. Grew and Moult were rivals in exploiting the waters at Acton and at Shooter's Hill. Francis Moult and his brother George, who was elected a Fellow of the Royal Society in 1689, were reputable apothecaries; Francis was a yeoman of the Apothecaries Society 1691–1719. (Hunter, 1982a p.72). This was admirably told by Dr Alex Sakula (1984) in his Gideon de Laune Lecture, and I am much indebted to him for further information.

Bridges in the preface to his translation of 1697 wrote that Grew 'had pitch'd upon a Citizen of London well reputed, who being perfectly instructed hath made the salt for some years in great quantities for the Shops and for Physicians for their private use'. This was Thomas Tramel, whom Grew recommended in a letter sent in May 1698 to various physicians, (his draft is in the British Library: Sloane

MS 4037,f.77) complaining of Moult's counterfeit and asking 'what you find in the use of [Grew's own salt] worthy of remarque I shall be glad to know; and with your leave shall insert it into a 2d edition'. This second edition of his Latin *Tractatus* came out later in 1698, when he also obtained a Patent for fourteen years' monopoly in his making of Epsom Salt. (Patent 354, 15 July 1698 for 'The Way of Making the Salt of the Purgeing Waters perfectly Fine, in Large Quantities and very Cheape, so as to be commonly Prescribed and Taken as a General Medicine'; the full title was kindly given me by Mr John Appleby from the *Specification of Patents 1693–1721*).

Bridge's translation was published again in 1700, but anonymously; this was almost certainly a second piracy by Moult, who was exposed in the vindication of Grew by Josiah Peter, enlarging his four-page 'Narrative' of 1697 into a sixty-page pamphlet *Truth in Opposition to ignorant and malicious Falshood* (1701).

This 'Vindication' begins by denying Grew's plagiarism of Malpighi's discoveries in botany, quoting praise of Grew from Malpighi himself and from Evelyn, Ray and the Frenchman Denis Dodart. Peter quoted letters from half-a-dozen European scientists written to Grew as Secretary of the Royal Society, which are still in the Society's archives. He mentioned the King's 'satisfaction' at Grew's support for Fitzgerald's distillation of sea-water. Coming to his main subject, Grew's *Bitter Purging Salt*, he printed many letters and 'testimonies' from physicians, apothecaries and 'eminent and learned persons' praising the product and exposing the danger of using the 'counterfeits' sold by George and Francis Moult, 'Johnson another chymist, and the two Sarfatis'.

There are two interesting letters to Grew from the Dublin physician Sir Patrick Dun describing the case of his patient the Duchess of Ormonde, for whom he had prescribed with ill effect the 'wrong' Epsom Salt bought in good faith from London. Peter mentioned Tramel as 'overseer of Dr Grew's salt works', and denounced Moult for printing 'as is supposed' 1500 copies of the English version of Grew's *Treatise*, and publishing an Advertisement in four newspapers 'that he still prepares and sells the Purging Salt', after Grew had obtained his Patent and given notice of it in the *London Gazette*. Grew had offered 'for peace sake' to allow the interlopers to use his method, but 'they showed no intention to come to any reasonable terms, except compelled by law'.

Peter quoted a 'testimony' of 1700 from Sir John Floyer, and I have found in the Bodleian copy of Bridges's 1697 edition of Grew's *Bitter Purging Salt* a contemporary note quoting Floyer's mention of 'The artificial bitter salt of Epsom' in his *Enquiry into Bathes in England* (also of 1697). Benjamin Allen in his *Natural History of the Chalybeat and Purging Waters of England* (1699) described (pp. 122–26) 'Ebbisham commonly called Epsom Water in Surrey' without mentioning Grew, but in his preface acknowledged that 'the learned Dr Grew recommended Epsom Waters, which my honoured friend and learned and compleat physician Dr Clopton Havers informed me sometimes succeed in the cure of a Diabetes'.

In his sixtieth year Grew published *Cosmologia Sacra*, a treatise expounding the support given to Christian belief by scientific knowledge, with the sub-title 'The Universe as it is the Creation of God'. Publication was recorded in the Term Catalogue for Trinity [June] 1701, Divinity Books no. 51 (TC III, 255); on 30 July Grew gave the Royal Society a copy of his book, which it still possesses.

He wrote in the Preface that 'The many leud opinions, especially those of the Antiscripturists published of late years by Spinosa and some others in Latin, Dutch and English, have been the occasion of my writing this Book'. Several of his contemporaries in the Royal society, who had made the deepest research into 'natural knowledge', testified that their discoveries, far from leading to scepticism, enhanced their religious faith. John Ray, whose botanical work ran parallel to Grew's, had published ten years earlier, in 1691, *The Wisdom of God in the Works of the Creation*, a book widely read which reached its fourth edition in 1704 and continued to be reprinted. *Cosmologia Sacra* seems to have aroused little comment, though Evelyn and Locke knew it; one hundred years later Coleridge annotated a copy, and in our own time Dr H.R. McAdoo, Archbishop of Dublin, in his book *The Spirit of Anglicanism* (1965) claimed *Cosmologia Sacra* as a good example of the Anglican theory of its day, though Grew was a Presbyterian.

In dedicating the book to King William III, Grew prayed that 'God may bless your Majesty's endeavours for the Redemption of the Foreign Churches', remembering his fellow Calvinists in the Low Countries and the Rhine Provinces occupied by Louis XIV since 1672, but he added an 'Epistle Dedicatory' to the two Archbishops of the established Church of England, Thomas Tenison of Canterbury and John Sharpe of York, who were tolerant of dissenters; in the reaction from James II's Roman Catholic bias the Church had accepted the Act of Toleration of 1689, which was promoted by Gilbert Burnet, Bishop of Salisbury. This toleration ended in December 1711, three months before Grew died, with the passing of the Conformity Act, against which Grew's friend and pastor John Shower had protested. (See Appendix III).

Where Ray had marshalled his profound knowledge of botany and zoology in evidence of the Creator's wisdom and power, Grew described the new interpretations of matter in the 'Corporeal World' and of mind in the 'Vital World', relating them to his conception of celestial mind and the laws of Providence in the government of the universe and of public states. He offered a firm basis for these interpretations in a long vindication of the historic truth of the Bible; he had adumbrated this argument long before in his discourse on *Roots* (1673, part 2, sections 1–7, reprinted in *The Anatomy of Plants* 1682, pp.74–81).

Grew attacked pantheism and scepticism by implication, avoiding advertisement of the opinions he deplored. He controverted a few earlier writers, dismissing for instance Hobbes's views on 'fancy' (p.23) and Willis's on 'vital fluid' (p.32), though both these writers had died more than twenty years before, as had Spinoza, rebuked by Grew (p.202) for denying an Old Testament miracle. On the other hand in defending Bible truth Grew quoted with approval the sixteenth-century Jesuit mathematician Juan Bautista Villalpando who had made the Biblical description of the Temple of Jerusalem accord with the principles of Vitruvian architecture; Villalpando had added an account of the City and Temple of Jerusalem to his edition of Jeronimo Prado's commentary on Ezekiel, published in Rome 1596–1605. (M.J. Ryan 'Villalpando' *Dictionary of Scientific Biography*, New York 1976, v.14, pp.29–30, discussing his connection with Galileo's school and his influence on Newton).

The range of Grew's interests and knowledge is clear from his chapter headings; it extended far beyond his particular studies. He wrote in the preface that he gathered his astronomical information from 'the best astronomers of the present age, having neither health, leisure nor convenience for nocturnal observations'; at the Royal Society he was in touch with Newton and Halley, and while Secretary in 1678 he corresponded with Jan Hevelius at Danzig, who regularly sent reports of astronomical observations made at various places in Europe. 'For the rest', Grew added 'so much only excepted as is historical, Nature has been my book'. The treatise comprises five 'books': I and II consider the material and psychological aspects of 'nature', III discusses natural and moral law, IV and V occupy two-thirds of the volume in defence of the historical truth of the Bible.

His survey of the Bible was historical, not theological; he had learned Hebrew 'the first occasion I ever had' to read the Old Testament in its original language.

Like other leading Fellows of the Royal Society, Grew accepted the atomic theory, and in discussing 'the formed atoms which are the principle of all matter' (Book I, chapter 3, paragraph 11) he mentioned the phenomenon of electricity. This had been known for half a century, but had been little considered except by Boyle, whose tract on electricity in *Experiments about the Mechanical Origin of Qualities* (1675) 'brought the term into common usage, and is the first work on electricity in the English language' (J.F. Fulton *Bibliography of Robert Boyle*, 2nd ed. 1962, no.123; for reference to earlier work see Fulton's pages 73 and 88).

In his chapters on biology Grew compared the tissues of animals and plants (Book II, chapter 2), a main subject of his own study, though he exaggerated his analogies. Chapters 3–6 on mind expressed a physician's view of psychology; he divided the faculties of mind as sense, fancy, and reason, discussing their interaction and the influence of the body on each. He also described how the fancy of one person may operate on that of another. (R. Hunter and I. Macalpine *Three hundred years of Psychiatry 1535–1860*, 1963, pp.285–87 quoting and discussing five paragraphs from *Cosmologia Sacra* Book II, chapter 6).

VII   THE ROYAL SOCIETY

VIIA   GREW AS SECRETARY AND EDITOR   1677–79

After Henry Oldenburg's death on 5 September 1677 the Royal Society appointed Robert Hooke and Nehemiah Grew joint Secretaries in his place. Among Grew's papers in the British Library is a draft in Latin of 'The Circular Letter sent to divers of ye forrein Correspondents' announcing that Lord Brouncker had retired from the Presidency 'through weariness of age' – 'statu corporis valetudinario laborans' –, and Sir Joseph Williamson had been elected in his place 'on the recent St Andrew's Day' 30 November 1677. Since Oldenburg its Secretary had died 'some months ago' the Society 'conferred the charge on us who have signed our names below. Correspondents are invited to address either of us as Secretary of the Royal Society at Gresham College'. The draft letter is signed by Grew and a second copy dated 31 December 1677 is countersigned by Robert Hooke. A third copy, slightly abbreviated, is addressed to Johann Hevelius of Danzig (Sloane MS 1942,f.1 & verso, with title on f.2 verso; second draft f.3 and verso: 'Londini pridie Kalendas Jan: 1677 – Robert Hooke'; third f.4 and verso; f.30 verso, a fourth partial draft of the circular), telling him that the Secretaryship has been conferred 'on Mr Robert Hook and me together' ('D. Rob^O Hook . . . Mihique una') and suggesting that a gift from 'his splendid and extensive collection, adorned with his name' would be most welcome to the Society. No gift from Hevelius is recorded in *Musaeum* (1681); he had been in controversy with Hooke since 1674, and in 1679 many of his astronomical instruments and other possessions were desroyed in a fire. His acknowledgement of the circular is among Grew's official letters at the Royal Society, dated from Danzig 18 June 1678; his Latin report of the occultation of Jupiter on 5 June 1679, dated from Danzig 10 June, is in Hooke's *Philosophical Collections* 1 (1679), 29–32.

Between February 1677/8 and February 1678/9 Grew edited six issues of the *Philosophical Transactions*, nos.137–142, and was, presumably, responsible for the General Index 'from the Beginning to July 1677' published during 1678; the six issues were paginated continuously as the concluding parts of Oldenburg's volume 12 though Grew was instructed by Wren on 4 May 1678 to suspend publication. (Sloane MS 1942, f.27, A.L.S. Sir Christopher Wren to Grew 4 May 1678; Wren as Vice-President had taken the chair at the Council meeting that day – Birch v.3, pp.406–08). He continued as Secretary till late in 1679 when, on 24 November, Hooke wrote to Newton 'Dr Grew's more urgent occasions having made him decline the holding correspondence, the Society hath devolved it upon me'. (Sir Isaac Newton *Correspondence* edited by H.W. Turnbull. Cambridge 1960, v.2, p.297: letter 235).

Hooke had already sent to the printer on 9 October the first number of his *Philosophical Collections* and continued the publication through seven issues, till the Society's Council on 13 December 1682 authorised revival of the *Transactions*, which began its volume 13 with no. 143 for January 1682/83. (Robert Hooke *Diary* edited by H.W. Robinson, London 1935, p.429; Sir Geoffrey Keynes *A Bibliography of Robert Hooke,* Oxford 1960, pp.46–50; *Philosophical Collections*, photographic facsimile, New York, Johnson and Kraus Reprint Corporations 1965).

In the issues which he edited Grew continued Oldenburg's method, printing articles and reports by Fellows and Correspondents, with a few 'accounts' of new books; he called himself 'Author of the Transactions', a description suggesting that he wrote these reviews of books. The account of Martin Lister's *Historia Animalium Angliae* in no.139 appears to be by Grew, perhaps also the notice of Hooke's *Lectures and Collections*, while Lorenzo Legati's *Museo Cospiano*, noticed in no.140, was used by him in compiling the Catalogue of the Society's Museum during these years 1678–79. He acknowledged his authorship of two pieces: the translation from Latin of his half-brother Henry Sampson's 'Anatomical Observations' in No.140 and the official Latin Letter to Leeuwenhoek in No.142. Much of No.141 is concerned with astronomy, of which Grew confessed himself ignorant, though 'relying on the best astronomers of his time' for information (*Cosmologia* 1701).

Leeuwenhoek's long correspondence with the Royal Society was described in detail more than fifty years ago by Clifford Dobell in his encyclopaedic survey of the pioneer microscopist's work *Antoni van Leeuwenhoek and his 'Little Animals'* (London 1932). Dobell found that the Dutch autograph letters in the Society's archives had been published only in abbreviated translations in the *Philosophical Transactions*; complete texts were printed in *Opuscula selecta Neerlandica* 9 (1931). During 1980–81 the autographs were freshly examined by Brian J. Ford who found viable microscopic specimens attached to the letters by Leeuwenhoek three hundred years ago: 'The van

Leeuwenhoek specimens sent to Oldenburg... 1674' *Notes and Records of the Royal Society* 36 (1981), 37–59; Professor Ford has further described 'The Leeuwenhoekiana of Clifford Dobell' in the same journal 41 (1986), 95–105, with perspicacious assessment of Dobell's work and character.

Grew was concerned, while Secretary, with the following letters from this correspondence:

> [1677] Two letters from Leeuwenhoek to Oldenburg *Phil. Trans.* 140,pp.1002–05.
>
> Nov. 1677 Leeuwenhoek to Brouncker *Phil. Trans.* 142,pp.1040–43.
>
> 1 Jany 1677/8 Grew to Leeuwenhoek *Phil. Trans.* p.1043.
>
> 18 March 1677/8 Leeuwenhoek to Grew *Phil. Trans.*, pp.1044–45.
>
> 31 May 1678 Leeuwenhoek to Grew *Phil. Trans.*, pp.1045–46.
>
> 27 Sept. 1678 Leeuwenhoek to Grew Dobell (1932) pp.179–81.
>
> 21 Feb. 1678/9 Leeuwenhoek to Grew (mentioned in the next letter)
>
> [Spring] 1679 Leeuwenhoek to Grew *Phil. Collections* No.1,pp.3–5.
>
> 25 April 1679 Leeuwenhoek to Grew (text not printed) Dobell (1932) p.184,n.3.

In the interval between publication of Grew's last issue early in 1679 and Hooke's first number in November the editor of the *Journal des Scavans* in Paris, J.B. de La Roque, wrote to Grew offering to publish reports from the Royal Society if the *Philosophical Transactions* were discontinued (Sloane ms 1942, f.6: English translation by Grew of letter from J.B. de La Roque, Paris 18 October [N.S. = 7 Oct. O.S.] 1679).

The first number of Hooke's *Philosophical Collections* was compiled before Grew resigned his secretaryship; it includes, besides Leeuwenhoek's letter to Grew, three unsigned notices of books, which I attribute to Grew:

The account (p.38) of Malpighi's *Anatome Plantarum* Part 2, published by the Royal Society during 1679, is merely a list of its contents with a eulogy of 'the most ingenious Author... the real Phosphor Bononiensis'; Malpighi had sent his manuscript to Grew as Secretary on 21 June 1678, and it is preserved among Grew's papers in the Sloane manuscripts, inscribed on the cover in Grew's hand 'Senior Malpighius's Discourse' (ms 1942, ff.15–26).

A more detailed summary (pp.39–42) of Denis Dodart *Mémoires pour servir à l'Histoire des Plantes* (second edition, Paris 1679) quotes in translation several passages about the chemical 'virtues' of plants which resemble Grew's own writings and use some of his technical terms; Grew wrote in the preface of *The Anatomy of Plants* (1682) that 'Not long after [the printing of his *Idea* in 1673] I received a curious and learned book from Mons. Dodart... Fellow of the Royal Academy at Paris, in pursuance of whose order it was published, which being a design of a like import I was glad to see it justify'd by that illustrious Society'. Dodart's *Mémoires* published in folio and octavo in 1676, represented the *Projet* of 'the whole Académie' whose report was not printed till 1731 (*Mémoires de l'Académie royale des Sciences*, v.4, pp.423–536.)

Nine passages (pp.42–43) are paraphrased 'for a specimen' of Stefano Lorenzini *Osservazioni intorno alle Torpedine*, Florence 1678; the last of these discusses the fish's 'Stomachs, Guts, &c.' as in Grew's lectures of 1677, while the review ends by praising the Author for his 'new comparative Anatomy', Grew's own method of research.

From time to time Grew served on special Committees, among them one to consider Hooke's demonstration in 1674 of Leeuwenhoek's discovery of protozoa, in 1678 one which found Moses Pitt's project for his great *Atlas* too grandiose, and in May 1679 'to discourse with Mr Cheney concerning Chelsea College' (Birch *History* v.2, pp.387, 480; v.3, pp.502); but he left no account of this work himself.

Grew's work for the Royal Society has been recorded by Dr Michael Hunter in four masterly appreciations, between 1981 and 1985 (see the list of References, p.xvi). Dr Hunter's discoveries and conclusions have helped me much, and I am indebted to him for further personal information. While he is primarily interested in Grew's work in and for the Royal Society, I have tried to portray Grew as observer, thinker and writer.

1673 : Observations touching the Nature of Snow.
1684 : The Pores in the Skin of the Hands and Feet.
1691 : Observationes ... de Morboso Liene.
1693 : A Query concerning the food of the Humming Bird.
1711 : The Number of Acres in England and the Use which may be made of it.

1   *Observations touching the Nature of Snow*   1673

Grew sent his 'Observations touching the Nature of Snow' in a letter to Henry Oldenburg dated March 12, 1671/2. The paper was read at the Royal Society on 4 April 1672 and published a year later, with slight abbreviation, in the *Philosophical Transactions* for 25 March 1673; the original text has been printed among Oldenburg's *Correspondence*.

The hexagonal form of snowflakes had been described by several observers before Grew, notably by Kepler (1611), Descartes (1637), and by both Boyle and Hooke in 1665; Dr and Mrs Hall conclude, in their notes on the original letter, that Grew did not know Kepler's work since he names only the other three predecessors. Dr Joseph Needham has recorded even earlier Chinese observations in his *Clerks and Craftsmen in China and the West*, 1970. The European writings were well summarised by D'Arcy Thompson in his famous book *On Growth & Form* (1917), but neither he nor Needham mentioned Grew's contribution.

Hoping to 'add a little to the great deal of light given to the nature of cold in [Boyle's] excellent discourses', Grew described the different forms of flakes observable 'in a thin, calm and still snow': stars of six points, irregular figures, broken fragments, and 'others by various winds gently thaw'd and then froze into clumpers again'. The discussion which he appended is a theoretical attempt to explain the chemistry of ice formations, and uses analogies such as 'the configuration of feathers ... from the urinous parts of the blood of fowls' to suggest that the 'little star' of a snowflake arises from the reaction of the spiritous particles in a drop of rain with the saline particles 'partly nitrous but chiefly urinous'.

2 The Pores in the Skin of the Hands and Feet 1684

This account of the pores and ridges in the skin of the hands and feet was published in the *Philosophical Transactions* on 20 May 1684. Grew had already mentioned them when describing the pores in the stems of plants in *The Comparative Anatomy of Trunks* (1675), where he wrote (p.48) 'The pores are so very large in the Trunks of some Plants that they are visible to a good eye without a glass, being so large as very well to resemble the pores in the skin of the fingers and ball of the hand'.

The pores in the skin had been known since antiquity, but Grew described those in the hands and feet as 'very remarkable in respect of their position and amplitude, hitherto described by no anatomist... The papillae, observed with an indifferent glass, are arranged in little ridges everywhere running parallel one with another, especially upon the ends and first joynts of the fingers and thumb, upon the top of the ball and near the root of the thumb a little above the wrist, disposed into spherical triangles and ellipticks... And so those of the feet... Upon these ridges stand the pores all in even rows... The use of the pores is for the discharge of the more noxious and perspirable parts of the blood'. He did not mention the tactile function of the skin-ridges which Malpighi had discussed in 1665 in his *De externo tactus organo*.

Edward Tyson in his *Anatomy of a Pygmie* (1699) noticed the 'lines' in the chimpanzee's hand, but Grew's observation was not taken further for nearly a century, when William Cruikshank published in 1795 a coloured engraving of the papillary ridges, followed in 1823 by Jan Purkinje's fuller discussion and more accurate illustration. Classification of 'finger-prints' was undertaken only at the end of the nineteenth century, and then applied to personal identification.

Dr Alex Sakula has kindly drawn my attention to the paper by M.J. Leadbetter 'Nehemiah Grew M.D., F.R.S., the first dactyloscopist?' *The Finger-print whorl'd* (Finger Print Society) April 1977, pp.57–58.

3 *Observationes de morboso Liene* 1691

A medical history and post-mortem report from Grew's practice, describing a rare disease of the spleen, was communicated to the Royal Society on 13 May 1691 and published in the autumn in Latin.

A daughter of Thomas Sedgwick, a London merchant, started to learn embroidery at the age of fourteen, and worked incessantly for two years. Her melancholy temperament turned to illness, with loss of colour and appetite, failure of menses, and cough. After three years pain settled low on the left side, and persisted till her death from fever at the beginning of her twentieth year. *Post mortem* all was found healthy except the spleen which was enlarged and tumid: more than 2 fingers thick, 4 broad and 10 long, weighing more than 25 ounces. A diseased spleen is usually hardened with dark scirrhous tumours; here the whole substance was putrescent and foetid, dissoluble by the palpating fingers, like grumous blood; the part not held broke off by its own weight; it was as red within as without, but there was no abscess nor truly purulent matter. *Conclusion*: The excessive size was casued by unequal distribution of aliment from prolonged lack of exercise, which is specially necessary in adolescence. The blood which ought to have been evacuated by menstruation was thrown back to the spleen. *Corollary*: In girls undue intermission of exercise, particularly around the end of their second seven years or the beginning of the third, is very bad.

This report gives no indication that Grew was aware of recent study of the spleen, as recorded by Lambert van Velthuysen in his *Tractatus duo ... de liene* 1657 or by Johann Conrad Brunner in his *Experimenta nova* 1682, in spite of his own interest in anatomical research.

4 *The Food of the Humming Bird* 1693

Grew communicated to the *Philosophical Transactions* for May 1693 a letter describing a humming-bird, evidently written by an Englishman who had visited North America. His own paper published in the issue for July and August, suggested that the bird collects insects from the flowers it visits, not nectar as the 'letter' stated.

The writer of the letter published in May is here identified as Mr Hamersley of Coventry; Grew questioned his statement that the bird sucks honey from the corolla of flowers and suggested that it picks up insects with its beak, which he contrasted with a bee's proboscis; he added that 'the bird should be opened'. Hamersley probably knew only the ruby-throated humming-bird of eastern North America (*Archilochus colubris*), which he described as of 'a shining green colour, something resembling our drake's heads'.

Humming-birds from South America were described in the sixteenth century, but the first English report seems to be in Thomas Morton's *New English Canaan . . . New England* (1637); Evelyn noted in his *Diary* on 11 July 1654 that he saw in the Anatomy School at Oxford 'two humming-birds not much bigger than our humble bee'; and Thomas Green mentioned them 'of divers colours' in his 'Account of Virginia' in the *Philosophical Transactions* 11 (1676), 631.

Grew himself had described (*Musaeum*, pp.61–63) two skins in the Royal Society's collection: one being 'the greater kind, above four and a half inches long . . . of a blackish-brown or Eagle colour . . . and no where radiant' [*Archilochus alexandra*?], and the other 'the lesser humming bird. His head is lost. From the top of his breast to the end of his tail two inches long, but the body alone not above $\frac{3}{4}$ of an inch . . . on his wings and tail a dark brown, on his belly a yellowish red, on his breast white, on his back green mixed with glorious golden rays' [*A. colubris*?]. Grew believed that the earlier naturalists had copied their descriptions from Oviedus (Juan Gonzales Oviedo *Historia general y natural de las Indias occidentales* 1535), but quoted their accounts of the bird's note and flight. Grew wrote that 'he feeds by thrusting his bill into a flower like a bee . . . John de Laet [Jan de Laet *Historia naturalis Brasiliae* 1648, includes Georg Markgraf's records] saith that his tongue is twice as long as his bill . . . and it is very like to be so, as a part more apt, by its length and flexibility, to wind itself to the bottoms of the deepest and most crooked flowers: in which the honey-dew which these birds, as well as bees, do suck is usually lodged'. With opportunity to dissect a humming-bird, Grew might have observed the quasi-tubular tongue with which it collects from flowers both nectar and insects. In this passage of *Musaeum* he noted that 'Marggravius describes nine sorts of them'; a century later Buffon (*Histoire naturelle – Oiseaux*, v.6, 1779) recorded forty-three species. Buffon's bird volumes were written by Philibert Guéneau de Montbéliard.

5   *The Number of Acres in England*   1711

This paper summarises an elaborate and methodical project for economic and social advance, which Grew had intended to submit to the Privy Council while the Union of England and Scotland was being negotiated in 1706; the summary was printed in the *Philosophical Transactions* in 1711. The manuscript of the project belonged to William Petty-Fitzmaurice, first Marquess of Lansdowne, who died in 1805; his collection was bought in 1807 for the British Museum, where Grew's paper seems to have remained unnoticed by economists till about fifty years ago. Lord Lansdowne was a great-grandson of Sir William Petty F.R.S., who had carried through the Cromwellian Survey of Ireland and died in 1687.

Grew, in his summary, criticises Petty's estimate of 28 m. acres for the area of England and Wales as 'computed not measured'; reckoning by triangulation 'where the wheel-measure is taken along the roads' he stated his result to be over 46 m. acres; the modern surveyed assessment is 38 million. Comparing this reckoning with the ratio of acreage to population (1 : 2.4) in the Netherlands, he concluded that England and Wales could support 110 million people and might well support half that number. The population when Grew wrote in 1706 was above five and a half million, had less than doubled to nine million in 1800, and reached his proposed 55 million only after more than two hundred years. This summary paper suggested that 'industry, inclosure, and improvement', meaning universal employment and efficient management of agriculture, crafts and trade, could double the population in twenty-five years and quadruple it in thirty-six. Two hundred ports could be developed and provide for more trade than all the ports of Europe. Similar improvement could later be achieved in Scotland and Ireland. Grew's reference to a wheel-measure recalls the 'Waywiser given by Bishop Wilkins' to the Royal Society, as recorded in *Musaeum* (1681) page 360.

The original project elaborated these suggestions in practical proposals for their realisation. In his preliminary address to Queen Anne he claimed to be proposing 'improvements which England is capable of, and Scotland too in some measure, if Your Majesty's glorious design [in her speech at the opening of Parliament] of uniting the Two Kingdoms shall be accomplished'. At the beginning of his main text he wrote 'I have discoursed with miners, farmers, artificers, merchants and others... I have also read above 40 authors', implying that his project was a distillation of widely-held opinions.

Among potential improvements he recommended a geological and mineral survey, metalled roads, canals dug by 'the engine invented by Mr Bayly which takes up a ton and a half in little more than a minute', enclosure of land leading to businesslike management of estates, development of vineyards and orchards, newer crops and better breed of animals. For manufactures and trade he proposed better ships and harbours, free trade under the law especially from company monopolies, more manufactures and exports, less 'mere selling trades' handling imported goods, encouragement of mechanical inventions such as Robert Holt's 'tool which draws steel into long pieces'. For increasing the population he looked to universal employment, 'blocking up ways' to useless and unproductive professions, immigration of skilled craftsmen, marriage inducements and a tax on bachelors, more equal care for health by regulating physicians' fees, hospitals for nursing mothers and prevention of excessive infant mortality among rich and poor.

He hoped for large improvement in existing habits of life and work without anticipating an industrial revolution, though he foresaw that the American colonies would seek independence. His project revived the social ideals of the Commonwealth years, rejected at the Restoration and later ridiculed in Swift's *A Voyage to Laputa* (*'Gulliver'* part 3, 1726) and similar satires, as we have seen social conscience sacrificed to mammon in the 1980s. Grew hoped to realise the ideals of his republican youth by deploying the intellectual approach of the Royal Society through scientific method and mechanical invention.

The manuscript was brought to notice by the American economist E.A.J. Johnson in the *American Economic Review*, vol.21, pp.463–80 and fully discussed in his *Predecessors of Adam Smith* (1937), chapter 7 'Nehemiah Grew the scientist'; Johnson concluded that Grew 'sketched a methodical picture of England's possibilities' and praised his 'strongly utilitarian treatment'. I learn from Dr Michael Hunter that Grew's project has been studied recently by Professor Neal Wood of York University, Toronto.

VIII ENQUIRIES RELATING TO NEW ENGLAND
AND THE INDIANS 1690

Two letters written to Grew from New England in the winter of
1690–91 answered questions which he had sent to Samuel Lee,
who had emigrated in 1686 and was pastor of the church at Bristol,
Rhode Island; Lee was a well-known dissenting theologian.
Grew's letter of questions about the institutions of Massachusetts
and the Indians in New England has not survived, but Lee's answer
and a covering letter from Samuel Sewall of Boston, which added
'some few animadverions of my own' are among Grew's papers in
the British Library; they were published by George L. Kittredge in
1913.

The answers in Lee's letter are numbered 1–118, evidently to
match Grew's questions. After describing mercantile and profes-
sional life, particularly the practice of medicine in the Colony, he
wrote 'as to Indians' about their medical, psychological and moral
condition. French pox and consumption were their worst diseases,
but 'the great arrow of God' (plague) was unknown, though
smallpox was very frequent. He described their herbal remedies,
their longevity, their mating with 'little or no love, almost like the
beasts', the occasional suicide of women when their men will not
acknowledge them or their children, the method of carrying children, the seclusion of girls at puberty. He was shocked by one
question of Grew's, presumably, Kittredge surmised, about
homosexual practices. Lee wrote of the Indians never weeping,
and their patience and courage under torture; the scalping of ene-
mies – some scalps had been sent to England; their amusements of
football, dancing and dice; their general abstemiousness and occasional drunkenness; their 'howling' at funerals. He had received
'most of the intelligence' from Dr Caleb Arnold of Portsmouth,
Rhode Island.

Sewall's letter described the local weather, and told how eager
he was to educate and convert the Indians; he thought that they
ought to live in reservations.

Many travellers had described and portrayed Amerindians since
the beginning of European settlement, but Grew's questionnaire

was possibly the earliest English search for a detailed anthropological survey. He published nothing from the answers he had obtained, though ten years earlier he had described a few Indian artefacts in the Royal Society's Museum: weapons, 'wampam-peage', clothing made of feathers, and 'an Indian pail made of the barque of a birch-tree' (*Musaeum* 1681, pp.364–74).

IX   Account of Henry Sampson   1705

Nehemiah Grew had two half-brothers Henry and William Sampson; his mother Helen, daughter of Gregory Vicars, had been married first to William Sampson, a Puritan poet. After Sampson's death in 1636, she married Obadiah Grew on Christmas Day that year; they had three children, of whom Nehemiah was the second. He was about twelve years younger than his half-brother Henry, but they became life-long friends.

When Nehemiah went up to Pembroke Hall, Cambridge in 1658 Henry was a Fellow of the College, lecturer on Greek philosophy, and Rector of Framlingham, Suffolk, a College living, though not episcopally ordained. He was ejected from his Rectory and Fellowship in 1662 under the Act of Uniformity, when the Church of England was reestablished after the King's restoration; he remained in Suffolk preaching in private houses but was prohibited from this ministry by the Conventicle Act of 1664. Henry then studied medicine in France, Italy and Switzerland, graduated M.D. at Leiden in 1668 and settled in the City of London as a physician. Henry encouraged Nehemiah in his botanical work and put him in touch with Oldenburg, Secretary of the Royal Society; no doubt his example encouraged Nehemiah to graduate at Leiden. They were elected Honorary Fellows of the Royal College of Physicians on the same day, 30 September 1680. Henry Sampson retired in 1699 to Clayworth, Nottinghamshire where his younger brother William, who conformed to the Church, had been Rector since 1662; Henry died there in 1700.

Grew wrote his 'Account of Henry Sampson' to supplement the 'Memorial' of his religious career by John Howe, the Presbyterian minister to whose congregation Henry had belonged in London; Howe published the 'Memorial' and the 'Account' in his devotional *Discourse of Future Blessedness* 1705. Howe's 'Memorial' recalled that Sampson had attended the same congregation for thirty years, read the Bible in Hebrew and Greek from his youth, and had formed a library of Bibles. He had died after 'a long and languishing distemper, suffered with a calm and composed temper'.

Grew's 'Account' added biographical facts, recording Sampson's outstanding career at Cambridge, his work as a Puritan minister, and his study of medicine at 'several foreign universities' before going to Padua and Leiden. Henry had left unfinished a record [now in the British Library department of Manuscripts] of 'Observations of Diseased Bodies dissected with his own Hand', besides historical writings.

Edmund Calamy used material collected by Sampson for his *Account of the ejected Ministers* (1702) and included an account of Sampson in his *Continuation* (1727). There is a memoir by A. Gordon in *DNB* 50 (1897); the entry in Innes Smith *English-speaking Students of Medicine at Leiden* (1932) corrects an error in the records. F.J. Cole (1944 p.365) described his work with J.J. Wepfer at Schaffhausen in 1667; Frances Packard's memoir of Henry Sampson is in *Suffolk Review* 4 (1977), 256–63.

X CORRESPONDENCE

Little of Grew's correspondence survives and few of the extant letters speak of his personal life; some have come down the years only because Grew wrote botanical notes on their blank backs. They provide a record, fragmentary indeed, of his busy life spent in research and its publication, in the affairs of the Royal Society, and in the practice of medicine, and show glimpses of his enquiring, friendly yet firm character ready to defend himself when criticised or plagiarised. Most interesting are the letters to and from Malpighi about their parallel researches in botany, which have been superbly edited by Howard B. Adelmann. A similar group between Grew and Martin Lister contains criticism and defence of Grew's published writings, but the controversy was about points of little import. In connection with both these series there are numerous letters concerning Grew in Oldenburg's *Correspondence*, excellently presented in Dr and Mrs Hall's vast and thorough edition. Oldenburg, as his habit was, kept these collaborations or confrontations in play for the interest of the Royal Society as a clearing house for scientific news; after his death in September 1677 letters from the Society's correspondents, still addressed to him through many subsequent months, were dealt with by Grew as Secretary till early in 1680. Twenty years later, between 1697 and 1700, Grew allowed Josiah Peter to use several of these official letters in his pamphlet *Truth in opposition to Falshood*, with others written personally to Grew and testimonials requested from prominent men by Peter, all in evidence of Grew's recognised distinction as a scientist and physician robbed by interloping apothecaries of his rights in Epsom salt. Among these is a history from the eminent Dublin physician Sir Patrick Dun of the severe illness of his patient the Duchess of Ormonde after taking 'counterfeit' salt (see above p.51). Michael Hunter (1982a, note 74) records a letter of 1682 from Grew to John Evelyn in the Evelyn collection at Christ Church, Oxford.

Some letters from Oldenburg and Hooke as well as Grew mention dispatch of his published books to Malpighi, and his own letters to Martin Lister discuss subscriptions to the *Musaeum* Catalogue and to *The Anatomy of Plants* from Lister's friends in Yorkshire; while the dedicatory letters in his works record the course of his research. The letters he obtained from New England about the Indians reflect an unexpected facet of his intelligent curiosity. We owe the preservation of the majority of these letters to the assiduity with which the early Secretaries of the Royal Society filed their papers, and not least to Hans Sloane who acquired Grew's miscellaneous notes and draft letters and such official papers as Grew had not deposited with the Society, incorporated them in his vast collection, and bequeathed them to the British Museum.

Grew died suddenly on 25 March 1712, aged seventy, at Racket Court, Shoe Lane. Racket Court is the second entry on the north side of Fleet Street coming west from the City across the now covered Fleet river. The *Prospectus* issued in 1680 for *Musaeum* named Grew's address as 'near Warwick Court in Warwick Lane', on the other (east) side of the Fleet and a little to the north; the College of Physicians had moved to Warwick Court in 1669. Grew was at Racket Court in 1691 when Samuel Sewall wrote to him there from Massachusetts; from July 1704 till 25 May 1711 he rented the Fellow's house on the north side of the college courtyard in Warwick Lane, with permission to drive his coach across the court if he kept the paving in repair (Clark *History of the R.C.P.* 2 (1966), 535, and information from Dennis Cole, Librarian R.C.P. 1982, also A. Sakula *Clio medica* 19 (1984), 4). I assume that he lived chiefly in Warwick Court and practised from Racket Court.

Grew's death was registered as of Christ Church parish, but he was a member of a Presbyterian congregation, of which John Shower was Minister from 1691 till 1715, first at Curriers Hall, London Wall and later in Old Jewry, where Shower preached Grew's funeral sermon. Shower recalled that during his own illness in 1706 he had been attended by George Howe, Francis Upton, Richard Blackmore and Nehemiah Grew. When he published the sermon he dedicated it to Blackmore, for the other two had also 'died in a sudden and surprising manner', Howe in 1710 and Upton in 1711.

After speaking of Grew's father and half-brother, both devout Presbyterians, he described Nehemiah's knowledge of history, his proficiency in Hebrew, his acquaintance with the theories of the heavenly bodies, skill in mechanics, mathematics and chemistry, in anatomy and medicine, and in the knowledge of plants and minerals. Grew's character he praised as grave and serious, yet affable and courteous, just and tolerant, kind and compassionate; 'those who knew him best believed he has the least mis-spent time to answer for of any they ever observed'.

John Shower was a younger brother of Sir Bartholomew Shower, Counsel for Robert Fitzgerald in the controversy with William Walcot in 1683 about their 'Sea-Water' patents.

The Bibliography

THE MANUSCRIPTS and printed editions of Grew's books and pamphlets are here described in chronological sequence of their first editions as set out in the preceding pages. Because it combines and completes the series of his writings on botany and chemistry *The Anatomy of Plants* 1682 is described before *Musaeum* 1681.

Representative copies are recorded from many libraries in these islands and in North America, where I have been generously allowed to examine many or they have been described for me by numerous libarians. I have looked particularly for copies inscribed by Grew or his contemporaries or owned by notable people, but have attempted no census of copies, except for a few scarce editions, nor have I enquired widely among European libraries.

Timothy Crist, while revising 'Wing' volume 2 at Yale, at my suggestion searched for variant issues of Grew's published books, which Donald Wing had inadvertently conflated, and kindly informed me. *NUC* also conflated variants; in quoting it I have abbreviated its Grew number 05115 to the two extra digits denoting a particular edition.

In setting out the contents of a volume I have accepted the recent advice of several bibliographic authorities to abbreviate the 'algebra', not keeping to the rigour of the game, but recording, for example, pagination without the corresponding signature references, where this gives adequate identification of a book's make-up, being concerned with the publishing history and Grew's occasional modifications of his terminology, without seeking the 'ideal text' of a literary composition.

Disputatio Medico-Physica, inauguralis, de Liquore nervoso. Quam, favente Deo Opt. Max. ex auctoritate Magnifici Dn. Rectoris, D. Abrahami Heidani, S.S. Th. Doct. ejusdemque Facultatis in Inclyta Academia Lugd. Bat. Professoris ordinarii, & Ecclesiae Pastoris disertissimi atque vigilantissimi, nec non Amplissimi Senatus Academici consensu, & Nobilissimae Facultatis Medicae decreto, pro Gradu Doctoratus, summisque in Medicina honoribus & privilegiis rite consequendis, Eruditorum examini subjicit Nehemias Grew, Anglus e Com. Warvvicensi. Die 14. Iulii, loco horisque solitis. [device] Lugduni Batavorum, apud Viduam & Haeredes Joannis Elsevirii, Academiae Typograph. [rule] cIↄ I ↄc LXXI

COLLATION: 4°: A⁴; 4 leaves, unnumbered. Page height 23 cm.

CONTENTS: A1ᵃ Title-page; a1ᵇ Dedication; A2ᵃ–4ᵇ text.

DEVICE: Elzevir's olive-tree, on the right a robed man, on the left a scroll inscribed NON SOLUS.

COPIES: Leiden State University Library 236 B5:30 without inscription; British Library 1185.g.11 (19); Royal College of Physicians of London 117.a.18(10); [U.S.] National Library of Medicine (NUC:NG 0511561).

Manuscript copy 1675

Disputatio Medico-Physica de Liquore Nervoso read Oct 28:75 [In a different contemporary hand:] *Dr Grew A 1672 julii 16* [recte 14 Julii 1671]. F°: 6 leaves, numbered in a modern hand 42–47; no dedication or introduction, 11 pages of text, f.47v blank. Royal Society CP XIV (i) Physiology, no.27; at one time folded in four, endorsed in a formal 17th-century hand *Grew de liquore Nervoso 72*; the paragraphs are not numbered. Birch *History of the Royal Society* 3: 228–29 records that it was ordered to be registered and Grew was desired to print it, but he did not find it in the Register.

II THE ANATOMY OF PLANTS

IIA THE ANATOMY OF VEGETABLES
BEGUN 1672

Manuscript

Anatomia Vegetalium Inchoata nec non Vegetationis Historia generalis abinde exacta. Autograph Latin draft, unfinished. British Library, Sloane MS 1926, ff.193, 194 and 197 rectos and versos; corrections for the English text, ff.190 recto, 191, 192, 195, 196 rectos and versos, 198 recto, 199 recto and verso.

Printed Editions

[In a double rule] The Anatomy of Vegetables Begun. With a general Account of Vegetation founded thereon [rule] By Nehemiah Grew, M.D. and Fellow of the Royal Society [rule] London, Printed for Spencer Hickman, Printer to the R. Society, at the Rose in S. Pauls Church-Yard, 1672.

COLLATION: 8⁰: A⁸, a⁸, B–N⁸, O⁷; 16 leaves, 186 pages, 1 blank leaf, 9 leaves; 3 folding plates. Page-height 148 mm.

CONTENTS: A1ᵇ [Royal Society's imprimatur] Novenb.9.1671; A2ᵃ Title; A3ᵃ [Dedication] To the Right Honourable & Most Illustrious The President & Fellows of the Royal Society; A4ᵃ–7ᵃ The Epistle Dedicatory To the Right Reverend John, Lord Bishop of Chester; A8ᵃ–a1ᵇ The Preface; a2ᵃ–7ᵃ The Contents; a7ᵇ–8ᵃ quotation from Glisson de Hepatis Anatomia; a8ᵇ To be added and corrected; pages 1–35 Chap. I Of the Seed as Vegetating; pp.36–65 II Of the Root; pp.66–98 III Of the Trunk; pp.99–103 An Appendix. Of Trunk-Rooots(!) and Claspers; pp.104–23 IV Of the Germen, Branch, and Leaf; pp.124–28 An Appendix. Of Thorns, Hairs and Globulets; pp.129–48 V Of the Flower; pp.149–67 VI Of the Fruit; pp.168–86 VII Of the Seed; N6 blank; N7ᵃ half-title The Explication of the Figures; N8ᵃ–07ᵃ Explications of Figures 1–29; Plates 1–3.

The figures are in broken sequence: Plate 1 carries figures 1–5, 15, 16, 19; plate 2-figs.6–8, 20–29; plate 3-figs.9–14, 17, 18; a note on the errata page states 'The Reader is desired to excuse the misplacing of the Figures by the Graver in the Authors absence.'. Plate marks (1) 8 × 9.5 cm; (2) 15 × 9 cm; (3) 15 × 3 cm.

MISNUMBERED PAGES: 50–51 as '48–49', 54–55 as '52–53', 58–59 as '56–57', 62–63 as '60–61', 178 as '187', 186 as '198'. Running title across openings of text: The Anatomy/of Vegetables.

Hickman the publisher presented four copies of the printed book at the meeting of the Royal Society on 7 December 1671, but the imprint is 1672, and it was listed in the Term Catalogue on 7 February 1671/2; priced Two Shillings.

TC I, 96; Wing G 1946; *NUC* – NG '54

COPIES: British Library 987.e.20; British Museum (Natural History); Linnean Society, inscribed *Ant. Swab/London 6s* and *E. Bibl. Linn 1784 J.E. Smith*; Royal Botanic Gardens, Kew; Royal College of Physicians inscribed *J. Merrick/Duplicate from St John's College Oxford*; Royal College of Surgeons; Royal Horticultural Society.
[Note on Inscriptions: *Anton von Swab* (1702–68) Swedish mineralogist, in London 1736. Linnaeus later persuaded Uppsala to buy Swab's important collection of minerals, and evidently acquired some of Swab's books for his own library, including this and Grew's *Trunks* (1675). (Details of Swab's career kindly given by C.O. von Sydow, Keeper of Manuscripts, Uppsala University Library). Sir James Smith bought Linnaeus's library with his herbarium, etc. in 1784 and founded the Linnean Society of London upon them in 1788. James Merrick, probably the poet and Fellow of Trinity College, Oxford (1720–69).]
Royal Society, inscribed *Presented from the Author to the Royal Society Dec, 7. 1671*; Royal Society of Medicine with a contemporary MS note quoting Velthusius *de Liene* 1657 on development of plants, and signature of *John Bostock* [vice-president of the Royal Society 1832]; Wellcome Institute. Cambridge University Library U^X.7.53 (E), and Sir Geoffrey Keynes copy, *Bibliotheca Bibliographici* 2461; Trinity College, Dublin; Royal Botanic Gardens, Edinburgh; Bodleian Library, Oxford – Lister A 218 inscribed *Bibliothecae Ashmoleanae dedit Martinus Lister M.D.;* Reading University Library, Cole 1969, no.782. Bibliothèque Nationale, Paris S 13517; Groningen University, Netherlands; Osler Library, Montreal *Bibliotheca Osleriana* 2837; National Library of Medicine, Bethesda [Washington D.C.]; *NUC* records seventeen further copies in the United States.

2 *Second Edition* 1682

The Anatomy of Plants, Begun. With a General Account of Vegetation, Grounded thereupon . . . The Second Edition. In *The Anatomy of Plants* 1682 leaves E4^a–N1^a, pages 25–49 'The First Book'.

3 *French translation* 1675

Anatomie des Plantes qui contient une Description exacte de leurs parties & de leurs usages, & qui fait voir comment elles se forment, & comment elles croissent. Traduite de l'anglois de Monsieur Grew de la Société Royale. [ornament] A Paris, chez Lambert Roulland, Imprimeur-Lib. Ord. de la Reyne, en la maison d'Antoine Vitre, rue du Foin, devant la Petite Porte des RR.PP. Mathurins. [rule] M.DC.LXXV. Avec Privilege du Roy.

COLLATION: 8^0: π^1, a^{12}, A–R^8, S^4, T^6; engraved leaf, 12 leaves, 215 pages, 6 leaves, 14 plates. Page height 14 cm.

CONTENTS: Engraved title-page; a1^a Title; a2^a–5^a Epitre à Monsieur des Marests, [signed] LeVasseur; a5^b–7^b Avis au Lecteur; a8$^{a–b}$ Lettre escrite par Mr Grew à l'Autheur de la Traduction, Nov.16/26 1674 [in Latin]; a9^a–12^b Preface de l'Autheur anglois; pages 1–215 Text; p.216 Colophon as title-imprint, but 'Reine' for title-page's 'Reyne'; T1^a–6^a Table des Chapitres et des principales Choses contenues en ce Livre; T6^b Privilege . . . le 22. Octobre 1675 [for ten years to] Sieur LeVasseur . . . Registré le dix-septième jour de Février 1665 [sic] . . . Achevé d'imprimer pour la première fois le 16 Novembre 1675.

ENGRAVED TITLE: Flora (?) holding three lilies in her right hand, seated in a room containing books and botanical equipment, on a draped curtain at the top ANATOMIE DES PLANTES; signed *F. Chauueau in et fec*; plate-mark 12 × 7 cm.

PLATES: inserted through the text on unnumbered leaves carrying fourteen figures; the two plates which carry, respectively, figures 3, 4, 5, 14 and figures 6, 7, 9, 11 are repeated at the place appropriate for each figure.

ENGRAVED HEAD-PIECES: At a2^a emblematic of botany for the dedicatory Epistle, and three scenes as chapter headings – (1) two men examining plants, for chapters I, p.1, IV p.113, VI p.169, (2) two gardeners for chapters II p.37 and V p.147, (3) two botanists with a microscope for chapters III p.69 and VII p.193; there are nine engraved cul-de-lampes of varying size.

Marginal notes are printed on pages 18, 24, 44, 46, 48, 76, ,80, 120, 210. Chapters I and VI are headed 'premier' and 'sixième', the other chapters have roman figures. The w in Grew's name on the title-page and on a3^b and a8^{ab} is from a variant alphabet.

COPIES: British Library 724.a.23, Sir Joseph Banks's copy, Dryander's Catalogue 1798, v.3, p.363; Wellcome Institute; Cambridge University Library, Sir Geoffrey Keynes's copy, *Bibliotheca Bibliographici* 2462, which he generously lent me shortly before his death; Trinity College, and Marsh's Library, Dublin; Bibliothèque Nationale, Paris; Boerhaave Museum, Leiden (State Library of the History of Natural Science and Medicine); *NUC* – NG'47 – records nine copies in the United States including National Library of Medicine, Bethesda (Washington) and the Hunt Library, Pittsburgh (Jane Quinly *Catalogue* 1 (1958), no.338 with full collation and reproduction of the engraved-title and title-page on plate XXV). Montesquieu's copy was offered for sale in 1971 by the Paris bookseller Paul Jammes.

NOTE: The identity of LeVasseur the translator is discussed above at page 14. The dedicatee was presumably the nephew of Colbert, Louis XIV's great Secretary of State: Nicolas Desmarests became the King's last Controller-General of Finance in 1708 but was removed by the Regent Orléans, 'sacrificed to public hatred' according to Voltaire, for his sumptuary policy intended to repair the waste of the old King's wars.

4   French translation: reissue   1679

Anatomie des Plantes [as 1675] ... Société Royale. Seconde Edition [flower] A Paris, chez Antoine Dezallier, rue S. Jacques à la Couronne d'Or [rule] M.DC.LXXIX. Avec Privilege du Roy.

Reissue of the sheets of 1675 with a cancel title-page and T5–6 replaced by a new half-sheet V.

COLLATION: As 1675, including the *Extrait du Privilege*, reprinted on V2^b.

COPIES: British Library 1471.a.10; Wellcome Institute, London; Bibliothèque Nationale, and Bibliothèque interuniversitaire de Médecine,

Paris. *NUC – NG'*48: National Library of Medicine, Washington; University of North Carolina, Chapel Hill; New York Botanical Gardens; Jefferson College, Philadelphia; Missouri Botanic Garden, St Louis; Yale Medical Historical Library.

NOTE: This reissue of Louis LeVasseur's translation of *The Anatomy of Vegetables Begun* was published in Paris in the same year as Guy Mesmin's translation of *Luctation* (see the Table at p.11). Both translations were published together at Leiden in 1685 and reissued there in 1691, with related works by N. Dedu and Robert Boyle. This Leiden collection was published in an Italian translation at Venice in 1763.

5   *Latin*   1678

Anatomiae Vegetabilium Primordia cum Generali Theoriâ Vegetationis Eidem superstructâ operâ Nehemiae Grew, M.D. & Societatis Regiae Socii. Ex Anglicâ in Latinam translata.

In *Miscellanea Curiosa sive Ephemeridum medico-physicarum Germanicarum* Academiae Naturae Curiosorum. Annus Octavus M.DC.LXXVIII. Vratislavae & Brigae, Sumptibus Collectorum, Typis Johannis Christophori Jacobi anno M.DC.LXXVIII. Appendix, pages 287–379 (Nn3ᵃ–Bbb1ᵃ).

CONTENTS: in Latin throughout: p.287 Half-title; 289 Dedication; 290–92 Letter to John Wilkins, Bishop of Chester; 293–95 Preface; 296 quotation from Glisson; 297–375 text, with caption title; 377–79 Explanation of the Figures; Plate comprising 29 figures, 168 × 260 mm.

RUNNING TITLE: Anatomiae Vegetabilium Caput I[–VII].

NOTE: The *Catalogus Librorum* in this volume lists this translation, with promise that translations of *Idea and Roots* and of *Anatomy of Trunks* will be published in the next volume, as they were, with a translation of *Mixture. Miscellanea Curiosa* began publication at Leipzig in 1670 and was continued from various cities; the early volumes were reprinted at Nuremberg in 1693. Grew wrote in the preface to *The Anatomy of Plants* (1682) 'by the Ingenious Collectors of the German Ephemerides both my First, Second and Third Books are all published in Latine, but their unskilful Interpreter doth often fail of the grammatical sense'.

COPY USED: Royal College of Surgeons of England (Nuremberg reprint). The *British Union Catalogue of Periodicals* provides a summary history of the various series of publications from the Imperial Charles-Leopold Academy of Naturalists, now the Deutsche Akademie der Naturforscher, with their changes of title and a record of holdings in British libraries.

6 *Anatomy of Vegetables Begun, with Luctation, etc.* French, 1685

Anatomie des Plantes, qui contient une Description exacte de leurs Parties & de leur Usages, & qui fait voir comment elles se forment, & comment elles croissent. Par Mr Nehemiah Grew, Secrétaire de la Société Royale d'Angleterre: Enrichie de Tailles douces. Et L'Ame des Plantes par Mr Dedu, Docteur en Médecine. Avec un Recueil d'Expériences & Observations curieuses faites par Mrs Grew & Boyle. Très curieux & utile aux Médecins à & (sic) ceux qui s'appliquent à la recherche de la Nature des Qualités & des Proprietez de toutes sortes de Corps. [ornament] A Leide Chez Pierre Vander Aa, Marchand-Libraire. MDCLXXXV.

COLLATION: 12⁰: ★⁴A–N¹², /², A–D¹², E⁶; 4 leaves, 310 pages, 3 leaves, 108 pages; 1 plate. Page height 135 mm.

CONTENTS: ★1ᵃ Engraved title-page; ★2ᵃ Title-page; ★3ᵃ–4ᵇ Avis au Lecteur; pp.1–223 Anatomie des Plantes; pp.224–34 Explication des Figures; folding plate faces p.224; pp.235–46 Table des Chapitres; p.247 Half-title: De L'Ame des Plantes. De leur Naissance, de leur Nourriture & de leurs Progrez. Essay de Physique. Par M. Dedu, Docteur en Médicine de la Faculté de Montpellier. [device] A Leide [rule] Chez Pierre Vander Aa, M DC LXXXV.; pp.249–310 De L'Ame des Plantes; N12ᵃ Half-title: Nehem. Grew & Rob. Boyle de la Société Royale de Londres & Docteurs de Médecine Recueil d'Expériences et Observations curieuses sur le Combat, qui precède du Mélange des Corps, Sur les Saveurs, et sur les Odeurs. / 1–2 Preface Du Sr Nehem. Grew; pp.1–62 Expériences du Combat, Qui Provient de l'affusion & du mélange des Corps.; p.63 Half-title: Expériences curieuses de l'Illustre Mr Boyle, sur les Saveurs et sur les Odeurs.; pp.65–83: Chapitre I Des Saveurs, pp.83–108 Chapitre II Des Odeurs.

The engraved title displays an allegorical group, with Apollo above, signed *A. Schoonebeck invenit et fecit 1684*; the folded plate, 12 × 16 cm, unsigned but paginated 224, reproduces Grew's 14 figures; there is a cul-de-lampe on p.62 at the end of Grew's *Combat*.

CHAPTER HEADINGS: *Anatomie des Plantes* as in 1675 but Epines for Espines in the Suite du 3me Chapitre. The marginal headings for Chap. VI are misprinted IV at pages 177, 179, 180 and V at page 195. *L'Ame des Plantes*: I (pp.249–58) without heading, II (pp.259–67) La Naissance de la Plante, III (pp.267–71) Solutions de quelques difficulies [sic] sur cette matière, IV [misprinted VI] (pp.272–93) De la nouriture [sic] de la Plante [including:] Expérience Première (p.280),... Seconde (pp.281–2),... Troisième (pp.282–93), V (pp.293–310) Le progrez de la Plante. *Combat*: I (pp.1–14), II (pp.15–34), III (pp.34–62). *Saveurs et Odeurs*: I (pp.65–83) Des Saveurs. Expériences I–XII, II (pp.83–108) Des Odeurs. Expériences I–XII.

RUNNING HEADINGS: (1) AVIS AU LECTEUR. (2) ANATOMIE/DES PLANTES. (3) [Explication – no heading.] (4) TABLE. (5) DE L'AME/DES PLANTES. (6) PREFACE. (7) Expériences curieuses/du mélange des corps. (8) Expériences sur les Saveurs. (9) Expériences sur les Odeurs (Heading begins at p.97 though the text began at p.83).

COPIES: British Library 987.b.10, *Jos. Banks* facsimile stamp on title verso; British Museum (Natural History); Cambridge University Library – Geoffrey Keynes copy, *Bibliotheca Bibliographici* 2463; Oxford, Bodleian Library Vet.B 3/256; Paris, Bibliothèque Nationale; Netherlands: Amsterdam, Leiden and Utrecht University Libraries, Wageningen Horticultural college Central Library. United States: Library of Congress; Yale Medical Historical Library – J.F. Fulton's copy, see his *Bibliography of Robert Boyle* 2nd ed. 1961, no.131; *NUC – NG'49* records four more copies.

Dedu's book had been published in 1682: *De L'Ame des Plantes. De leur Naissance, de leur Nourriture & de leur Progrez: Essay de Physique.* Par M. Dedu, Docteur en Médecine de la Faculté de Montpellier. A Paris, Chez Estienne Michallet, rue S. Jacques, a l'Image S. Paul, M.DC.LXXXII. 12°: 3 leaves, 66 pages.

COPIES: British Library; Paris: Bibliothèque Nationale (Catalogue 36 (1908), S 13580), and Bibliothèque interuniversitaire de Médecine.

NOTE: My enquiries in France have found no record of Dedu's career; Vander Aa's second edition (Leiden 1691) describes him as Sieur N. Dedu; his publisher Estienne Michallet had published Mesmin's translation of Grew's *Luctation* in 1679.

7   *Anatomy of Vegetables Begun, with Luctation, etc.*   French, 1691

Anatomie et Ame des Plantes, qui contient une Description exacte de leurs Naissances, Nourritures, Progrez, parties & usages, & qui fait voir comment elles se forment, & comment elles croissent. Avec un Recueil d'Expériences & Observations curieuses, Par les Srs. N. Grew, R. Boyle, & N. Dedu. Enrichie de Tailles-douces. Nouvelle Edition, reveue & corrigée. [ornament] A Leide, Chez Pierre Vander Aa, [rule] M DC XCI.
   Reissue of the sheets of 1685.

COLLATION AND CONTENTS: as 1685, except for cancel title-page and half-titles reset in black and red; the *Avis au Lecteur* also reset in a different type, omitting Dedu's name; the Preface, inserted after N 12 in 1685, omitted here. The engraved title-page also omitted.

COPIES: British Library 987.b.11, *Jos. Banks* facsimile stamp on title-page verso; Edinburgh – Royal Botanic Gardens; Paris – Bibliothèque Nationale. The copy at the British Museum (Natural History) was destroyed during World War II. United Stated (*NUC* – NG '50): New York Botanic Garden; Williamsburg, Virginia – College of William and Mary. Though published at Leiden, no copy is recorded in the Netherlands Central Catalogue of Books. Fulton 1961, no.132 BL copy.

8   *Anatomy of Vegetables Begun, with Luctation, etc.*   Italian, 1763

Anotomia, ed Anima delle Piante, che contiene Una esatta Descrizione della loro origine, nodrimento, progresso, parti, ed usi, e che da a divedere come si formino, e come crescano. Con una Raccolta Di Sperienze, ed Osservazioni curiose sopra il combattimento risultante dalla mescolanza de' corpi, come anche sopra i Sapori, e sopra gli Odori, de' Signori N. Grew, R. Boyle, & N. Dedu. Tradotta dalla Lingua Francese, ed arricchita di figure in rame. [long rule/short rule] In Venezia, MDCCLXIII. Per Luigi Pavini. Con Licenza de' Superiori, e Privilegio.

COLLATION: 12°: A–N¹², o⁶; 320 pages, 2 leaves of Pavini's book-list; 1 folded plate (14 figures), plate mark 12 × 16 cm. Page height 163 cm.

CONTENTS: Title-page; pp.3–4 Avvertimento; p.5 half-title: Anotomia delle Piante; pp.7–156 Grew's text; pp.157–68 Spiegazione delle Figure;

pp.183–224 Dedu's text: Dell' Anima delle Piante; p.225 half-title: Raccolta di Sperienze . . . sopra il Combattimento . . . di Nehem. Grew e Rob. Boyle . . .; pp.227–78 Grew's text; p.279 half-title: Sopra i Sapori e . . . Odori; pp.281–320 Boyle's text.

COPIES: British Museum (Natural History) Botany Library; *NUC* – NG '46: Countway Library, Boston; Indiana University, Lilly Library; Missouri Botanic Gardens, St Louis; Yale Medical Historical Library (Fulton copy).

NOTE: The unsigned *Avvertimento al' Lettore* mentions 'Three curious Treatises written in the last century [i.e. Grew's *Anatomy of Vegetables Begun* and *Luctation*, and Dedu's *L'Ame des Plantes*] fallen into my hands. I have added Boyle's Experiments'. John F. Fulton (1961, nos.134 & 135) considered that this 'Advertisement' implied a previous, though untraced, Italian edition, but I believe it refers to one of the Leiden editions, because the title-page states 'Translated from the French'; G.A. Pritzel in his *Thesaurus literaturae botanicae* (1851) attributed this version to F.M. Nigrisoli.

IIB   AN IDEA OF A PHYTOLOGICAL
HISTORY   1673

Manuscripts

Historia Naturalis PLANTARUM Idea. Autograph draft in Latin with many corrections. Fo.: 11 leaves. British Library: Sloane MS 1926, ff.191–201.

Anatomiae Radicum comparatae Prosecutio, pars 2da. Autograph fair draft in Latin. Fo.: 38 leaves. The same MS, ff.144–79 and 187–88.

Miscellaneous botanical notes in Latin and English. Autograph. Fo.: 7 leaves. The same MS, ff.180–86.

First two paragraphs of *Idea*. Draft in Latin. Sloane MS 1941 f.84, probably autograph, but the hand more formal than in MS 1926.

Printed Editions

1   An Idea of a Phytological History Propounded. Together with a Continuation of the Anatomy of Vegetables, particularly prosecuted

upon ROOTS. And an Account of the vegetation of Roots Grounded chiefly thereupon. [rule] By Nehemiah Grew M.D. and Fellow of the Royal Society. [double rule] LONDON, Printed by J.M. for Richard Chiswell at the Rose and Crown in St. Pauls Church-yard, 1673.

COLLATION: 8^0: A^8, a^4, B–M^8; 12 leaves, 144 pages, 16 leaves; 7 folding plates. Page height 185 mm.

CONTENTS: A1 blank; $A2^a$ title-page; $A3–4^b$ The Epistle Dedicatory, to William Lord Viscount Brouncker the President, and to the rest of the Fellows of the Royal Society; $A5^a–7^b$ The Preface; $A8^a–a4^a$ The Contents; pages 1–53 An Idea of a Phytological History propounded, The First Part; pp.54–97 The Comparative Anatomy of Roots prosecuted. The Second Part. Chap. I[–V]; pp.98–144 An Account of the Vegetation of Roots grounded chiefly upon the foregoing Anatomy. THIRD PART; $L1^a–M8^b$ The Explication of the Figures; Plates 1–7.

The printer was John Martin.

Part 2 – 'The Anatomy of Roots' – is divided into five chapters without caption headings; the seven Plates carry 63 figures, numbered for each Plate, but without captions. The page headings run across the openings.

Wing G 1951; Term Catalogue I, 140 (16 June 1673, price 4s).
COPIES: British Library 449.c.2, and 987.f.4(3), Sir Joseph Banks's copy (Dryander's *Catalogue* 3 (1798), p.363); British Museum (Natural History), inscribed on flyleaf *London Mch.1 1707/8 H.D.* [Henry Dodwell, 1641–1711?]; Linnean Society R673; Royal Society of Medicine, lacking Imprimatur and Plate 7; Wellcome Institute, London. Cambridge University Library $Dd^x.3.35^1$(E) and Geoffrey Keynes copy, *Bibliotheca Bibliographici* 2465. Norwich: John Innes Institute. Oxford: Bodleian Library 8.a.66 Art, and Lister A 221 with inscription as in *Anatomy of Vegetables Begun*, and marginal notes in Lister's hand. Dublin: Royal College of Physicians of Ireland, lacking Plate 7, and Trinity College (2 copies). Edinburgh: Royal Botanic Garden, and Edinburgh University Library. Paris: Bibliothèque Nationale. North America (*NUC* – NG '68): Montreal, McGill University, *Bibliotheca Osleriana* 2838; Washington D.C.: Library of Congress, and Folger Library; Ann Arbor, University of Michigan; Cambridge, Massachusetts (Harvard) Arnold Arboretum; New York Botanical Garden; Philadelphia, University of

Pennsylvania, Botany Library; St Louis, Missouri Botanical Garden.

NOTE: John Locke (1632–1704) owned a copy – J.R. Harrison, *Library of John Locke* 1965, no.1322.

2 *Idea, with Anatomy & Vegetation of Roots*   Latin, 1680

Idea Historiae Phytologicae cum continuatione Anatomiae Vegetabilium, Speciatim in Radicibus et Theoriâ Vegetationis Radicum eidem praecipue superstructâ, proposita à Nehemia Grew, M.D. & Societatis Regiae Socio. Ex Anglicâ in Latinam translata.

   In *Miscellanea Curiosa* [see II A5] Volumes 9–10 (1678–79), 1680, Appendix, pp.99–218, with 7 plates.
   Reprinted Nuremberg 1693.

COLLATION: P.99 Half-title; pp.101–03 [Dedicatio]; pp.104–07 Praefatio; pp.108–42 Idea... Pars prima; pp.143–70 Comparativa Anatomia in Radicibus continuata, Pars secunda; pp.171–200 Theoria Vegetationis Radicum, fundata praecipue super praecedenti Anatomiae. Pars tertia; pp.201–18 Explicatio figurarum. Plates I–VI [and VI bis].

COPIES: As at II A5.

PLATES: Comprise 60 figures: I (42), II (11), III, IV, V (2 each), VI (1); Plate VI is in two forms – normal printing, and white on black.

NOTE ON TEXT: Grew's disparagement of this series of translations, in the Preface to *The Anatomy of Plants* 1682, is quoted above at *Anatomy of Vegetables Begun*, Latin 1678. He recorded in the same preface that Malpighi had procured a Latin translation of *Idea* (1673) to be made for his private use; no manuscript of such a translation is listed among Malpighi's surviving papers by C.Frati *Bibliografia Malpighiana* 1897, or H.B.Adelmann *Marcello Malpighi* 1966, pp.66–68, who described the loss of many of Malpighi's papers by his descendants, and gave a list of the present collections of manuscripts (pp.xvii–xix).

3 *Idea, with... Roots*   Second edition, 1682

An Idea of a Philosophical History of Plants. Read before the Royal Society, January 8. and January 15.1672... The Second Edition.

In *The Anatomy of Plants* 1682, leaves [1–3], pp. 1–24; 63 paragraphs.

The Anatomy of Roots; Presented to the Royal Society at several times, in the Years, 1672 & 1673. With an Account of the Vegetation of Roots, Grounded chiefly hereupon. The Second Book... The Second Edition.

In *The Anatomy of Plants* 1682 pp. 51–96. In Part I each chapter's paragraphs are numbered: 18, 7, 30, 23, 16 respectively; Part II has 70 paragraphs.

NOTE: The logical separation in 1682 of 'An Idea' from 'The Anatomy of Roots, necessitated renumbering Parts II and III of the original edition as Parts I and II of 'Roots'. The texts were only inessentially revised, with a few important changes of terminology:

1673	1682
Bubbles	Cells or Bladders
Lymphaeducts	Muciducts
Phytological	Philosophical
Succiferous	Lignous
Succiferous Vessels	Sap Vessels
Vegetables	Plants

In 'An Idea' paragraphs 1 and 2 were enlarged and no. 30 rewritten and enlarged; paragraphs 4, 6, 36, and 48–52 were added. In 'The Anatomy of Roots' a comparison of plant and animal tissues was dropped from Chapter IV para. 7, while a similar comparison was inserted into Chapter V para. 4. In 'The Vegetation of Roots' para. 37 reference to 'magnetic power' and 'electral Nature' was omitted. Chapter headings and marginal plate-references were added in 'The Anatomy'; and in 'The Vegetation' there are marginal subject-headings and references back to 'The Anatomy' besides plate-references. The seven plates of 1672 were replaced by thirteen larger plates V–XVII carrying 68 figures, with twenty 'standard circumferences' (gauges) on Plate XI.

IIC A DISCOURSE CONCERNING MIXTURE 1675

Manuscript

A Discourse Made before the Royal Society December 10. 1674. concerning the Nature, Causes, and Power of Mixture. By Dr

Nehemiah Grew. 2^0: 24 pages, not autograph. Royal Society: RB4, 271–85.

Printed Editions

1 A DISCOURSE Made before the ROYAL SOCIETY, Decemb.10.1674. Concerning the Nature, Causes and Power of MIXTURE. [rule] By Nehemiah Grew, M.D. and Fellow of the R. Society. [rule] LONDON, Printed for John Martyn Printer to the Royal Society, and are to be Sold at the Bell in St Pauls Churchyard, 1675.

COLLATION: 12^0: π^{10}, B–F^{12}; 1 blank leaf, 9 leaves, 120 pages. Page height 138 mm.

CONTENTS: $\pi2^b$ Royal Society's imprimatur January 21.1674/5; 3^a title-page; 4^a–10^b Epistle Dedicatory 'To the Right Honourable William Lord Viscount Brouncker President of the Royal Society'; Pages 1–120 text: introductory summary (pp.1–4) and SECT. I–V.
   Section V includes 'Instances' I–VI. No chapter headings, but running headings above the Epistle Dedicatory; pagination of the text in square brackets centrally in upper margin. Woodcut headpieces, with monkeys and human demi-figures, for Dedication and Text, also woodcut text initials O and H.

Wing G1948; Term Catalogue I, 205 – 10 May 1675.
COPIES: British Library 1651/1134, formerly in Patent Office Library; Cambridge, Magdalene College – Pepys Library (*Catalogue of Printed Books* 1978, no.97.2); Oxford, Bodleian Library (2 copies) Ashmole A43, and 8^0.Y.7 Art, in contemporary blind-tooled calf after Sir William Petty *The Discourse of . . . duplicate Proportion . . . in elastique Motion* 1674; Paris, Bibliothèque Nationale S13521. United Stated (*NUC*–NG '62): Library of Congress; Harvard – Arnold Arboretum; University of Wisconsin, Madison; Yale Medical Historical Library.
   Further copies are recorded with *Luctation* combined issue 1678, II E2 below.

2 *Mixture*   Latin, 1680

Discursus Habitus coram Societate Regia, Decemb.10.1674. Concernens Naturam, Causas et Vires Mixtionis, à Nehemia Grew,

M.D., & Regiae Societatis Socio. Ex Anglicà in Latinam translatus. In *Miscellanea Curiosa* Vols 9–10 (for 1678–79), 1680 Appendix.

CONTENTS: Page 295 half-title; pp. 297–300 dedication to Lord Brouncker; 301–26 text.

COPY USED: Royal College of Surgeons of England (1693 reprint).

NOTE: Fuller account of this series of translations is given under *Anatomy of Vegetables Begun*, Latin 1678, II A5.

3 *Mixture* Second Edition, 1682

A Discourse Read before the Royal Society Decemb.10.1674. Concerning the Nature, Causes, and Power of Mixture.

In *The Anatomy of Plants* 1682: P.217 Half-title: Several Lectures Read before the Royal Society.; p.218 The Titles of the following Lectures. – I. Of the Nature, Causes, and Power of Mixture. The second Edition; pp.219–20 Dedication; pp.221–36 Text; p.237 An Appendex to the precedent discourse of Mixture.

In the 'second edition' (1682) Grew called the Sections 'Chapters' and gave them headings; I. Of the received Doctrine of Mixture; II. Of the Principles of Bodies; III. Of the Nature of Mixture; IV Of the Causes of Mixture; V. Of the Power and Use of Mixture; the running headings correspond to these chapter-headings, but the numbering of this Lecture as part of the volume is misprinted 'Book IV' on pages 221–24, becoming correctly 'Lect.I.' only at page 225. 'Instances' III and IV in chapter V were combined as 'Instance III & IV'. The brief *Appendix* reported further experiments, and welcomed Denis Papin's 'Digester', which vindicated Grew's suggestion of 1675 that a combination of methods would accelerate mixture.

IID THE COMPARATIVE ANATOMY OF
TRUNKS 1675

Manuscript

A Generall and Comparative Description of the Stalke or Branch of a Plant. Presented to the R. Society by Dr Nehemiah Grew &c. [25 Feb. 1675]. Fo.: 25 pages, not autograph. Royal Society: RB4, 324–39.

The text corresponds closely to the first part of the printed text: the 'Anatomy of Trunks' without the 'Account of their Vegetation'.

Printed Editions

1   The Comparative Anatomy of Trunks, Together with an Account of their Vegetation grounded thereupon; in Two Parts: The former read before the Royal Society, Feb.25.1674/5; the latter, June 17. 1675. The whole explicated by Several Figures in Nineteen Copper-Plates; presented to the Royal Society in the years 1673 and 1674. [rule] By Nehemiah Grew, M.D. and Fellow of the Royal Society. [rule] London, Printed by J.M. for Walter Kettilby at the Sign of the Bishops Head in S. Pauls Church-yard. 1675.

COLLATION: 8^0: π^1, A^8, a^4, $B-G^8$, H^4; 13 leaves, pp.1–81, 11 leaves, 19 plates. Page height 183 mm.

CONTENTS: $\pi 1^b$ Royal Society's imprimatur, Octob.21.1675; $A1^a$ titlepage; $A2^a-5^a$ Epistle Dedicatory to Charles II; $A6^a-a1^a$ Dedication to Lord Brouncker and the FF.R.S.; $a2^a-3^b$ The Contents; $a4^a$ Errata; pages 1–81 Text: The Comparative Anatomy of Trunks, chapters 1–4; An Account of the Vegetation of Trunks, introduction and 7 chapters; The Explication of the Plates; Plates I–XIX.

The Explication provides the Roman numbering of the Plates, which carry Figure numbers only; I shows figs 1–7, II–XV show one figure each (8–21), XVI–XVII two figures each (22–23 and 24–25), XVIII, fig.26 is printed white on black, XIX, fig.27, not described in the Explication. Figures 1–21 show the dissected stems of seven herbs and fourteen trees, 22 'a Lactiferous Plant', 23 The Lymphaeducts, 24 a parcel of Air-vessels, 25 the Weftage of the Parenchyma and Vessels, 26 Lesser common Thistle, 27 shows a vertical rod, having ten rows of three rings and nine of two rings, irregulaly lettered A – Q.

The printer was John Martyn.

Wing G1947 Term Catalogue 1,220 (24 November 1675).
COPIES: British Library (3 copies) – 972.a.10 with many manuscript corrections, possibly in Grew's hand, largely but not completely used in the 'Second Edition' 1682; 987.f.3 Sir Joseph Banks's copy; Eve.a.77 John Evelyn's copy with his ownership inscription and press-mark M 15, and on the end paste-down brief notes of pages related to his own

writings; Banks's copy alone has the imprimatur leaf; British Museum
(Natural History); Linnean Society, inscribed twice *Ant Swab 6d/London*
(see note on their copy of *Anatomy of Vegetables Begun* 1672); Royal
College of Surgeons of England; Royal Society of Medicine. Cam-
bridge University Library U^x.6.82 (E), and Geoffrey Keynes copy,
Bibliotheca Bibliographici 2464; Edinburgh University Library; Oxford,
Bodleian Library (2 copies) – Lister A220 inscribed *Bibliothecae Ashmoleanae dedit Martinus Lister M.D.*, and Douce G443 with Francis
Douce's armorial bookplate; Oxford, Taylor Institution, Fry Collec-
tion from Upton Manor near Didcot, Herbals no.48 with signature and
printed label of Richard Hooper, Vicar of Upton, who helped Caroline
Sophia Fry (1861–1946) to form the collection (D.J. Gilson *Book Collec-
tor* 22 (1973), 47 & 56–57). Netherlands: Groningen and Leiden University Libraries. North America (*NUC* – NG '58): Montreal, McGill
University – *Bibliotheca Osleriana* 2839; Library of Congress and Folger
Library, Washington D.C.; Harvard, Houghton Library and Arnold
Arboretum; University of Wisconsin, Madison; New York Academy
of Medicine; New York Botanic Garden; Oberlin College, Ohio; Lib-
rary Company of Philadelphia, inscribed *Thomas Preston LCP 1789*; Yale
University Library.

2   *Trunks*   Latin, 1680

Comparativa Anatomia Truncorum Unâ cum Theoriâ Vegetationis Eorum eidem superstructâ in Duabus Partibus. Prior lecta
coram Societate Regiâ, Febr. 25.1674/5. posterior, Jun. 17.1675.
Tota explicata singulis figuris in novendecim tabulis aeneis; praesen-
tata Societati Annis 1673. & 1674. per Nehemiam Grew, M.D. &
Societatis Regiae Socium Ex Anglicâ in Latinam translata.

In *Miscellanea Curiosa* [as *Anatomy of Vegetables Begun*, Latin]
Volumes 9–10 (1678–79), 1680 (Reprinted: Nuremberg 1693), Appendix, with Plates VII–XVI.

CONTENTS: P.219 half-title; pp.221–25 dedication to the King; pp.226–
30 dedication to Lord Brouncker; pp.231–54 Comparativa Anatomia
Truncorum Pars I; pp.255–80 Theoria Vegetationis Truncorum fundata super praegressa Anatomia Pars II; pp.281–93 Explicatio Figurarum; 10 plates comprising 26 figures, mostly two to a plate.

COPIES: As listed at *Anatomy of Vegetables Begun*, Latin 1678, II A5.

3   *Trunks*   Second Edition, 1682

The Anatomy of Trunks, With an Account of their Vegetation Grounded thereupon. [rule] The Figures hereunto belonging, Presented to the Royal Society in the Years, 1673 & 1674. [rule] The Third Book. [rule] By Nehemiah Grew M.D. Fellow of the Royal Society, and of the College of Physicians. [rule] The Second Edition. [rule] London, Printed by W. Rawlins, 1682.

In *The Anatomy of Plants* 1682, pp.97–140, Plates 18–40.

IIE   EXPERIMENTS IN CONSORT OF
LUCTATION

Manuscript

Of the Luctation arising upon the Mixture of Bodies. By Dr Nehemiah Grew. Read before the R. Society April 13th 1676. 2⁰: 15 pages, not autograph. Royal Society: RB5, 136–46.

A continuation of the Discourse concerning Luctation arising from the Mixture of Bodies. By Dr Nehemiah Grew. Read before the Society June 1st 1676. 2⁰: 13 pages, not autograph. Royal Society: RB5, 147–56.

The first Discourse formed Chapters I and II of the printed edition, and the second formed Chapter III; the texts were revised for publication.

Printed Editions

1   *Luctation*   First issue, 1678

Experiments in Consort of the Luctation arising from the Affusion of several Menstruums Upon all sorts of Bodies, Exhibited to the Royal Society, April 13 and June 1.1676. [rule] By Nehemiah Grew, M.D. and Fellow the the Royal Society. [double rule] London, Printed for John Martyn, Printer to the Royal Society, at the Bell in S. Pauls Church-yard. 1678.

COLLATION: 12⁰; A⁶ B–F¹²; 6 leaves, 118 pages, 1 blank leaf. Page height 13 cm.

CONTENTS: A1^b Royal Society's Imprimatur, *Thursday, Novemb.* 15.1677.; A2^a title-page; A3^a Dedication to the Royal Society; A4^a–6^b The Preface; pp.1–118 Text; pp.1–26 Chap. I, pp.27–63 Chap. II; pp.64–118 Chap. III; F12 blank.

RUNNING HEADINGS: A4^a–6^b The Preface; pagination of text in brackets centrally at head of page.

Wing G 1950. British Library has no copy, though one was inadvertently recorded in Wing (information kindly given me by the Department of Printed Books 1987).

COPIES: Royal Society of Medicine, inscribed '*Duplicate. John Chatto. 1856*' (Chatto was then Librarian of the Royal College of Surgeons of England, which does not have this issue); Aberdeen University Library; Cambridge University Library: Syn 8.67.3; Oxford, Bodleian Library (2 copies) Lister A187 with Martin Lister's MS notes, and 8H.29.Med. in contemporary binding with Culpeper *Blood-letting* 1672 and Motte le Vayeur *Prerogative of private life* 1678; Oxford, Magdalen College; Paris, Bibliothèque Nationale S13522; United States (*NUC*–NG '65): Arnold Arboretum, Harvard; Massachusetts Historical Society, Boston; William A. Clark and Huntington Libraries, Los Angeles; Missouri Botanical Garden, St Louis; Yale University Library.

2   *Luctation and Mixture*   Second issue, 1678

Experiments in Consort of the Luctation arising From the Affusion of several Menstruums upon all sorts of Bodies. To which is added The Nature, Causes and Power of Mixture. Exhibited to the Royal Society. [rule] By Nehemiah Grew, M.D. and Fellow of the Royal Society. [double rule] London, Printed for John Martyn, Printer to the Royal Society, at the Bell in S. Pauls Church-yard, 1678.

COLLATION: As *Luctation* first issue and *Mixture* (1675), re-issued together, with a general title-page usually cancelling the first-issue title-page of *Luctation*, but retaining the title-page of *Mixture* dated 1675.

Recorded in the Michaelmas Term Catalogue, (T C, I.292) 26 November 1677 'Price 1s6d. Both sold by J. Martyn'; Wing first ed. and *NUC* do not differentiate the two issues; Wing 2nd ed. G 1950A.

COPIES: Royal College of Physicians of London with unsigned comtemporary notes on last fly-leaf; Royal College of Surgeons of England with the general title-page and the first-issue title-page of *Luctation*;

Wellcome Institute, London; Yale Medical Historical Library – J.F. Fulton's copy.

3   *Luctation*   Second Edition, 1682

Experiments in Consort of the Luctation Arising from the Affusion of several Menstruums upon all sorts of Bodies, Exhibited to the Royal Society, April 13. and June 1.1676.

In *The Anatomy of Plants* 1682, pages 238–54, as Lecture II of 'Several Lectures Read before the Royal Society'.

4   *Luctation*   Facsimile edition, 1962

Experiments in Consort... as original issue of 1678.

COLLATION: π^1, A^6, B–F^{12}; 7 leaves, 118 pages. Page height 135 mm.

CONTENTS: Verso of flyleaf: HEFFER SCIENTIFIC REPRINT; $\pi 1^a$–A1^a [Introduction headed] Grew's Experiments of Luctation, [signed at end] M.A. Hoskin; A1^b Royal Society imprimatur in facsimile; A2^a–F11^b facsimile of Dedication, Preface and Text.

Imitation parchment binding lettered up the spine GREW EXPERIMENTS OF LUCTATION, in a slip case.

Without date, published by Heffer, Cambridge 1962 and listed in *British National Bibliography*, January 1963, B63–3606 (class 615.1), Price 42s.

COPIES: British Library, etc.; Utrecht University Library; United States: *NUC 1956–67* records twelve copies.

5   *Luctation*   French, 1679

Receuil d'Expériences et Observations sur le Combat, Qui procède du melange des corps. Sur les Saveurs, sur les Odeurs, sur le Sang, sur le Lait, &c. Très-curieux & utile aux Médecins & à ceux qui s'appliquent à la recherche de la Nature, des Qualitez & des Proprietez de toutes sortes de Corps. [flower] A Paris, Chez Estienne Michallet, rue S. Jacques, de l'Image S. Paul. [rule] M.DC.LXXIX. Avec Privilège du Roy.

COLLATION: 12°; π^1, a^6, e^2, A–Y^6; engraved title, 8 leaves, 262 pages, 1 leaf; [plate] Fig. 1–5 facing p.229. Page height 157 mm.

CONTENTS: π1^a engraved title; a1^a title-page; a2^a–4^a Avis au Lecteur; a5^a half-title; a6^a–e2^a Préface de l'Autheur Anglois; pp.1–124 text of Grew's *Luctation*; p.1 Chapitre I, p.29 Chapitre II, p.59 marginal notes round outer and lower margin in small italic, p.68 Chapitre III; L3^a– T2^b, pp.125–220 Boyle's text, p.125 half-title; pp.127–67 Chapitre I. Des Saveurs; pp.168–220 Chapitre II. Des Odeurs; pp.221–62 Observations Faites avec le Microscope sur le Sang et sur le Lait, et communiquées a Mr Oldenbourg, Secrétaire de la Societé Royale de Londres, Par Mr Leuvvenhoeck de Delft en Hollande; P.221 half-title; pp.223–62 text of Leeuwenhoek's letters – Avril, Juin, Juillet 1674, Aoust 1675, Février 1698, pp.254–56 note in italic referring to Hooke's observations on the same subject, pp.256–9 further observations by Leeuwenhoek on salt, and pp.259–62 on manna; plate facing p.229; p.[263] Approbation de Messieurs les Doyen & Docteurs de Médecine de la Faculté de Paris... 4 Mars 1679; p.[264] Extraite du Privilège du Roy 10 Mars 1679..., Registré... 19 Avril 1679 [to Estienne Michallet for 6 years], Achevé d'imprimer le 18 May 1679.

Translated into French from Grew's English text, and presumably from Boyle's English and Leeuwenhoek's Latin also, by Guy Mesmin. Fulton records, nos. 130 and 133, Mesmin's Paris editions of 1679 and 1698, which comprise Grew on *Luctation*, Boyle on *Tastes and Odours*, and Leeuwenhoek on *Blood, Milk, etc.*, and also, nos. 131 and 132, the Leiden French editions of 1685 and 1691 which comprise LeVasseur's version of Grew *Anatomy of Vegetables Begun*, Mesmin's versions of *Luctation* and of Boyle *Tastes and Odours* with Dedu *Ame des Plantes*.

Misprints: p.89 numbered '8'; heading of p.137 'Saveues'.

Engraved title-page, unsigned: Allegorical figure [Flora?] seated, left hand pointing to shelves bearing books and retorts etc., right hand pointing to a paper lettered *Nouvelles Experiences*; in foreground plants, flowers and butterfly.

There is a two-line capital for the opening word of each chapter of the texts. Head-pieces of ornaments on leaves a2^a and 6^a and pp.169 and 223; woodcut head-piece with 2 putti, p.127; tail-pieces (3 tulip blooms) pp.124 and 153, (5 blooms) pp.167 and 220; ornament of 4 printer's flowers p.145.

COPIES: British Library 7407.a.34, contemporary signature *Ed. Simon* on title-page; Royal College of Surgeons of England, price *1–6* [1s.6d] inscribed on fly-leaf, bookplate of Robert Cony M.D. (1644–1713), lacks engraved title; Wellcome Institute; Bodleian Library, Oxford 8°

M 225 Th.; Reading University Library, Cole 1969 no.783; Paris –
Bibliothèque Nationale (2 copies); Bibliothèque interuniversitaire de
Médecine; Uppsala University, Waller collection. North America
(*NUC* – NG '77): National Library of Medicine; Berkeley; Cornell;
Wisconsin University, Madison; Yale Medical Historical Library (Ful-
ton copy).

6   *Luctation*   French, 1698

Recueil d'Expériences et Observations sur le Combat [as 1679...]
[ornament] A Paris, Chez Barthélemy Girin, a l'entrée du Quay des
Augustins, de coté du Port Saint Michel, à la Prudence. [rule]
M.DC.XCVIII. Avec Privilege du Roy.

Re-issue of the sheets of 1679 with a new title-page; collation and
contents as in 1679, including the Approbation, Privilege and Regis-
tration of that year; the engraved title-page of 1679 and the Plate to
face p.229 are not present in the British Library copy (1034.c.27), the
only copy I have been able to find, and apparently the only one
known to Fulton (no.133). In the first line of the title-page *Observations* is set irregularly with space after the letters O, B and E.

IIF   ANATOMY OF PLANTS   1682

1   *Prospectus*

A folded sheet, forming two leaves printed on all four sides;
302 × 194 mm.

COLLATION: 1ᵃ At a Meeting of the Council of the R. Society, Febr.
22th, 1681/2. [two paragraphs] Chr. Wren P.R.S. [rule across page] At
a Meeting of the R. Society, March 15th, 1681/2 Proposals made... for
Printing... The Anatomy of Plants, &c. by Dr Nehemiah Grew, in
one Volume... [8 paragraphs]; 1ᵇ Hereto are added several Lectures.
Read before the R. Society, by the same Author... [16 paragraphs]; 2ᵃ
[3 paragraphs, concluding the Prospectus] [rule across page] A list of
those Persons who have already subscribed... [104 names in 4 columns, the surnames aligned down the second and fourth columns, in

97

approximate alphabetic order]; 2^b [Conclusion of the list of names].

NOTE: The paragraph following each of the two dated headings on p.1^a begins with a three-line capital.

COPY: British Library: Bagford Collection of Title-pages, etc (former Harleian MS 5946) ff.164–65. I list the subscribers in Appendix I.

2 *Anatomy of Plants* 1682

[In a double rule:] The Anatomy of Plants. With an Idea of a Philosophical History of Plants. And several other Lectures, Read before the Royal Society. [rule] by Nehemiah Grew, M.D. Fellow of the Royal Society, and of the College of Physicians. [rule] Printed by W. Rawlins, for the Author, 1682.

COLLATION: fo.: π^4, a^4, B–Ii^4, Kk^2, Ll–Xx^4, Yy–Zz^2, Aaa–Ccc^2, Tab. 1–83; 11 leaves, 24 pages, 5 leaves, 304 pages, 10 leaves, 83 engraved plates. Page height 345 mm, large paper copies 365 mm.

CONTENTS: π1^b Royal Society imprimatur: Feb.22.1681/2, *Signed* CHR. WREN P.R.S., 2^a Title-page; 3^a–4^a Dedication: To his most Sacred Majesty Charles II. King of Great Britain, &c.; a1^a–4^b The Preface; B1^a half-title: An Idea of a Philosophical History... The Second Edition; B2^a Dedication: To The Most Illustrious The Royal Society,... and, In their Names also proposed to the Consideration of other Learned Men; B3$^{a–b}$ The Contents; pp.1–24 Idea; E4^a half-title: The Anatomy of Plants, Begun... The First Book... The Second Edition; F1^a Dedication: To the Right Honourable William Lord Vi-Count Brouncker, The President, And to the Council and Fellows of The Royal Society; F2$^{a–b}$ Epistle Dedicatory: To the Right Reverend John, Lord Bishop of. Chester... Coventry June 10.1671; F3^a–4^b The Contents; G1^a–N1^a, pp.1–49 text of The Anatomy of Plants, Begun.; N2^a half-title: The Anatomy of Roots... The Second Book... The Second Edition; N3$^{a–b}$ Dedication: To Lord Brouncker... September 1.1673; N4$^{a–b}$ The Contents; O1^a–S4^b, pp.57–96 text of Roots; T1^a half-title: The Anatomy of Trunks... The Third Book... The Second Edition.; T2$^{a–b}$ Dedication; To Lord Brouncker... President, and the Council and Fellows of the Royal Society ... London, August 20.1675.; T3$^{a–b}$ The Contents; pp.103–40 text of Trunks; Aa3^a half-title: The Anatomy of Leaves, Flowers, Fruits and Seeds. In Four Parts. The fourth Book;

Aa3[b] The Contents. Of the First Part; Aa4[a–b] Dedication: To The Honourable Robert Boyle Esq; pp.145–60 text of Leaves ... Read before the Royal Society, Octob. 26. 1676; Dd1[a] half-title: Flowers... Novemb. 9. 1676... The Second Part; Dd1[b] The Contents of the Second Part; pp.163–76 text of The Anatomy of Flowers; Ff1[a] half-title: Fruits... Read... in the year 1677. The Third Part; Ff1[b] The Contents of the Third Part; pp.179–92 text of Fruits; Hh1[a] half-title: Seeds... The figures presented... in the Year 1677. The Fourth Part; Hh1[b] The Contents of the Fourth Part; pages 195–212 text of Seeds; Ll1[a] half-title: Several Lectures Read before the ROYAL SOCIETY.; Ll1[b] The Titles of the following Lectures: I–VII; Ll2[a–b] Dedication to Lord Brouncker P.R.S.; pp.221–37, A Discourse Read., Decemb. 10.1674, Concerning... Mixture; pp.238–54 Experiments... of Luctation... Exhibited... April 13, and June 1.1676; pp.255–60 An Essay... Lixivial Salt... in Plants. Read... March 1676; pp.261–68 Discourse... Essential and Marine Salts of Plants. Read December 21.1676; pp.269–78 A Discourse of the Colours of Plants... Read May 3.1677; pp.279–93 A Discourse of... Tasts(!) Chiefly in Plants. Read... March 25.1675; p.294 Tabula, quâ perspicuè videre est, quot Triplicati Sapores, ex solummodo decem Simplicibus numerantur.; p.295 Tabula, quae Genericas omnes Saporum differentias conprenhedit(!); pp.296–304, Experiments in Consort upon the Solution of Salts in Water. Read... January 18.1676/7.; Yy1[a]–Aaa1[a] An Index of the Chief Matters...; Aaa1[b]–Ccc2[a] The Explication of the Tables...

PLATES: Tab.1–83; 14–17 and 40 are broad, folded plates 30 cm square; the average size of the rest is approximately 23.5 × 18 cm; 55 plates are numbered with Roman figures, 28 with Arabic figures; there are occasional variations of numbering in different copies, e.g. plate 14 sometimes lacks its number, or plate 15 is numbered 'Tab.6'. Most plates carry a heading description, and the individual drawings are named and usually numbered, with or without 'f' preceding the Arabic number. The separate books of 1672–75 comprised only 29 plates, mostly small; these were superseded in this volume by new engravings, except for figures 8–21 in *The Comparative Anatomy of Trunks*, which were retouched to form plates 22–35 in 1682; plates 41–83 are new. None of the plates are signed; they were presumably made from Grew's own drawings. Delineations, draughts, figures, or schemes were variously recorded in the Royal Society's manuscript Journal as shown by Grew at his Discourses on Pith and on Roots in 1672, on Trunks in 1674 and

1675, and on Flowers in 1676 and 1677; in his first Lecture on Flowers he remarked that his descriptions would be 'better conceived by showing some examples, the figures of which I have here present'; but when the Secretary entered the text of the discourse in the Society's Register he added in the margin 'These figures Dr Grew did not now Deliver to be here Inserted'. (*See* Plates 3, 4, 5, 6.)

Kk is a complete half-sheet, with correct catchword for Ll1 which follows; the gap in pagination ($Kk2^b = p.212$, $Ll1^a = p.[217]$) probably occurred because Ll was paginated in anticipation that Kk would be a full sheet; Kk ends the main text, Ll begins the 'Several Lectures'. Ll1 & 2 are not paginated, the numbering beginning with the text of Lecture I at Ll3^a, page 221. This was first pointed out by Joseph Bennett of the Lilly Library, Indiana University, Bloomington (*Biology* Exhibition Catalogue 1970, no.17, and n. at p.28).

ERRATA: p.150 for VI read IV; p.171 for Savilian read Sedleian; pp.221–24 running heading, for Book IV read Lect. I; leaf Tt2 is missigned Tt3; p.295 for comprenhedit read comprehendit; p.305 for Anotmy read Anatomy.

Not in Term Catalogue. Wing G 1945.
COPIES: British Library (three copies): 36.h.18 LP; L 35/86; 449.k.10; British Museum (Natural History); Chelsea Physic Garden; Linnean Society; Royal Pharmaceutical Society; Royal Botanic Gardens, Kew; Royal College of Physicians; Royal College of Surgeons of England, inscribed *Thos. Mauleverer* [1646–1701]; Royal Horticultural Society, Lindley Library; Royal Society LP; Royal Society of Medicine; University of London; Wellcome Institute. Bangor: University College of North Wales, Botany Department: title-page reproduced in Thornton & Tully *Scientific Books* 3rd ed. 1971 plate 7. Cambridge: University Library L.i.36 Bishop John Moore's copy, *Munificentia Regia 1715* bookplate; Botany School; Magdalene College, Pepysian Library (*Catalogue* 1978, no.2618) with Samuel Pepy's bookplate; Trinity College. Dublin: National Botanic Gardens, Glasnevin; Royal College of Surgeons of Ireland; Trinity College. Edinburgh: Royal Botanic Gardens; Royal College of Physicians, presented by the Author; University Library. Eton College: LP, early eighteenth-century College bookplate on title-page verso. Glasgow: University Library, William Hunter's copy. Norwich: John Innes Institute. Oxford: Bodleian Library (3 copies) – Sherard 651 LP, with Oxford Physic Garden bookplate; Lister D.20; C 1.25 Med. Seld.; and Radcliffe Science Library. Reading: Uni-

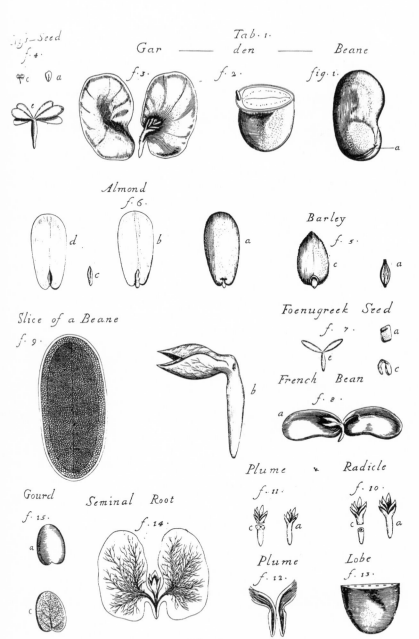

Seed
f. 4.

c   a

Tab. 1.

Gar ———— den ———— Beane

f. 3.   f. 2.   fig. 1.

a

Almond
f. 6.

d   b   a   Barley
f. 5.
c   c   a

Slice of a Beane
f. 9.

b   Foenugreek Seed
f. 7.   a
e
c

French Bean
f. 8.
a

Plume   ×   Radicle
f. 11.   f. 10.
c   a   c   a

Gourd   Seminal Root
f. 15.   f. 14.
a   Plume   Lobe
f. 12.   f. 13.
c

PLATE 3   *The Anatomy of Plants*
Plate 1: The Seed in its Vegetation

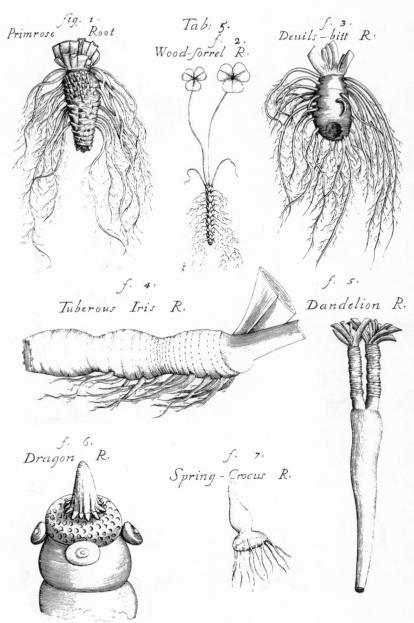

Primrose *fig. 1.* Root

Tab: 5.

f. 2.
Wood-sorrel R.

f. 3.
Deuils-bitt R.

f. 4.
Tuberous Iris R.

f. 5.
Dandelion R.

f. 6.
Dragon R.

f. 7.
Spring-Crocus R.

PLATE 4 *The Anatomy of Plants*
Plate 5: The Generation of Roots

Tab 18 Stalks & Branches cut transuersly

Indian Wheat

f.1
Dandelyon

f.2

Borage f.3

4 Holyoak

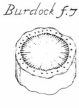

Colewort f.5

f.6
Wild Cucumer

Burdock f.7

Scorzonera f.8

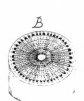

Endiue f.9

Vine f.10.A

B

f.11.Sumach

PLATE 5 *The Anatomy of Plants*
Plate 18: The Trunks of Plants cut transversely

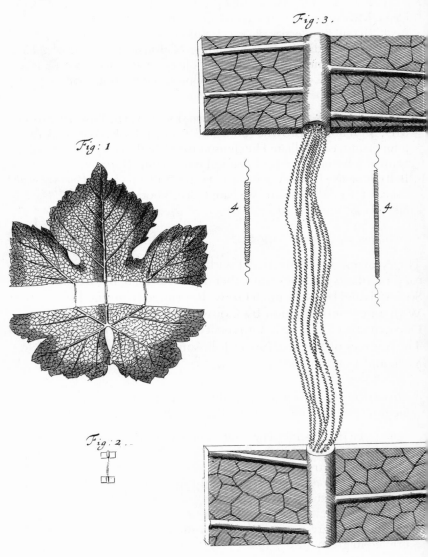

The Aer=Vessels unroaved in a Vine Leafe.

Fig: 3.

Fig: 1

Fig: 2.

PLATE 6 *The Anatomy of Plants*
Plate 51: Air Vessels in a Vine-leaf

104

versity Library, Cole 1969 no.785, LP. Paris: Bibliothèque Nationale – 2 copies (one LP); Amsterdam: University Library; Leiden: Boerhaave Museum; Wageningen: Horticultural College. Canada: Montreal: McGill University *Bibliotheca Osleriana* 2836. United States: *NUC* – NG'52 records 45 copies, including National Library of Medicine; Lilly Library, Indiana University; College of Physicians of Philadelphia; Library Company of Philadelphia; Hunt Botanical Institute, Pittsburg.

FORMER OWNERS OF COPIES NOT AT PRESENT TRACED: Edward Browne MD, Sir Thomas Browne (both sold 1711); John Evelyn (sold 1977); John Flamsteed; William Hutchinson 1686 (Zeitlin & VerBrugge 1964); Samuel Johnson (used for botanical terms in his *Dictionary* 1755); John Bell *Circulating Library Catalogue* 1775 and William Curtis *London Botanic Garden* 1783 (W. Noblett 'William Curtis's botanical Library' *The Library* 9 (1987), 1–22).

3 *Anatomy of Plants* 1965

The Anatomy of Plants [rule] with An Idea of a Philosophical History of Plants and Several other Lectures read before the Royal Society [rule] by Nehemiah Grew. Reprinted from the 1682 edition With a new introduction by Conway Zirkle Professor of Botany, Department of Biology, University of Pennsylvania, Philadelphia The Sources of Science, No.11 Johnson Reprint Corporation New York and London 1965.

COLLATION: Royal 8°: xix introductory pages; reduced photographic facsimile of the edition of 1682. Page height 28 cm.

CONTENTS: [i] half-title: The Sources of Science Number 11; [ii–iii] [List of] The Sources of Science Nos.1–14; [v] half-title: The Anatomy of Plants; [vi] Portrait of Nehemiah Grew reproduced from the frontispiece of *Cosmologia Sacra* 1701; [vii] title-page; [viii] copyright statement; ix–xviii Introduction; [xix] half-title, as on [v]; facsimile of 1682 edition, beginning with Imprimatur on verso of [xix].

Bound in green paper boards with green cloth spine; wrapper printed in green; price $35.00.

COPIES: British Library; Library of Congress; etc.

4   *Anatomy of Plants*   1967

The Anatomy of Plants... Microprint edition in 'Landmarks of Science' series, edited by I. Bernard Cohen and others. Totowa, New York: Rowman and Littlefield 1967.

Micro-photographic reproduction of the edition of 1682, on cards.

NUC–NG '53; *NUC 1965 through 1967* records this 'Landmark' series.

11G   ANATOMY OF PLANTS – RELATED PAPERS

'Dr Grew's Botanical Papers'. British Library: Sloane MS 2145.

A quarto–size volume of 295 leaves, comprising lists of plant-names in English and Latin and other memoranda in Grew's hand, written on paper of various sizes, the larger sheets folded and bound sideways. Some notes are on the blank backs of letters addressed to Grew during 1709.

The principal lists are:

ff.1–37: [Alphabetical list of plant names, English and Latin, summed by the writer to 1595 items.]

ff.42–48: 'A Catalogue of what I have in Stock with Latine & English Names.'

ff.49–55: '110 more Fruits & Seeds unknowne – In all about 720.'

ff.56–58v: 'A Catalogue of Seeds yet wanting or most of them.'

f.58v: 'Seeds to be procured.'

ff.63–90: [miscellaneous lists and notes].

ff.93–295v: [lists of exotic plants, with references to books of voyages; lists from [Ray's] 'Historia Plantarum &c'; lists of pears and apples, grasses, and other classes of plants arranged by genera with indication of their places of origin.]

Among the letters used as spare paper are two of botanical

interest:

ff.91–92: from Hans Sloane, 5 August 1709 'I am now putting in order my Jamaica fruits & so soon as they are rang'd you shall have notice.'

ff.60–61: from Robert Uvedale, 15 August 1709, discussing gardening and plants, thanking for a gift, offering seeds, and sending specimens of dried tulips 'of the year 1708 and one or two of this season'. Uvedale adds: 'allowance must be made for the fading of ye colours in dry specimens wch I could never yett find any trick to preserve in their native beauty.' (Uvedale (1642–1722) was a prominent cultivator of exotic plants. Sloane acquired his herbarium from his widow; it is described at length by Dandy (see below) 1950 pp.223–26.)

f.66: 'At Mr Thomas Sutton an Apothecary in Foster Lane may be had these Balsames at ye prices sett, vidt. [list of ten varieties priced from o.6s.od to o.1s.6d.]'

f.90: Draft in Grew's hand: 'Memorandm. An Indian Voyage. That Mr Scott is desired by Dr Grew to do him the favour to procure him of such Fruit Stones Berrys or Seeds as come more easily to hand ... And wt charge shall accrew shall be thankfully repayd. Given Mr Scott Dec. 7.1710 [with a list of four] 'Woods about a foot long & about an Inch & $\frac{1}{2}$ diameter.'

J.E. Dandy, ed. *The Sloane Herbarium* ... by James Britten, revised, London, British Museum (Natural History) 1958, Part 2 Biographical Accounts of contributors: p.132 'GREW (Nehemiah) ... His collection of fruits and seeds was bought by Sloane and is incorporated in the collection of "Vegetables and Vegetable Substances" but [Dandy added in error] there is nothing related to these in the Sloane manuscripts'.

L. Stewart 'The edge of utility: slaves and smallpox in the early eighteenth century' *Medical History* 29 (1985), 54–70 (especially 60–62) described Sloane's reliance on medical men stationed abroad and on agents of the Chartered Trading Companies for exotic plant specimens, chiefly between 1721 and 1735. Grew had made similar correspondence earlier.

Other papers written by Grew, but not directly connected with his publications, are among Sloane Manuscripts: MS 1941 for which see Appendix III; MS 1943 Prescriptions; MS 1945 Greek and Latin synonyms (1656); MS 1949 Observationes medicae (after 1679); MS 1950 Verses and Phrases (undated); MS 1963 ff.1–47 Egypt, ff.53–101 Catalogue of Books (c.1701).

Manuscripts

The British Library possesses Grew's autograph fair draft of the first six sections of his Catalogue:

Musaeum Regalis Societatis in Collegio Greshamensi. Latin, in ink on paper; 42 leaves, written on both sides. Folio. Sloane MS 1927 (Russell 333, note).

CONTENTS: f.1ᵃ *Caption title* – Lib. I De Animalibus, sect. 1 De Humanis. Mummia Aegyptica; f.3ᵇ: Sect. 2 De Quadrupedis; f.12ᵇ: Sect. 3 De Serpentibus; f.14ᵃ: Sect. 4 De Avibus; f.21ᵇ: Sect. 5 De Piscibus; f.33ᵃ: Sect. 6 De Testibus; f.39ᵇ–42ᵃ: Schema 1[–8]; f.42ᵇ: Ex his Schematis [Explanation – 7 examples, ending with Ostraea Gaderopeda].

The 83 pages of text record in Latin the specimens described, with more detail, in English on pages 1–153 of the published volume.

The British Library has two papers of extracts from the published *Musaeum*, by James Petiver and by Edward Lhwyd:

A Catalogue of the chiefest rarityes in Gresham College by Nahtnial [!] Grew D M. 81. English, in ink on paper: 4 pages. Sloane MS 2346.

CONTENTS: ff.21ᵇ–23ᵃ, in James Petiver's 'Medical and scientific Miscellany'. Seventy-one numbered entries, describing animal and bird specimens selected from *Musaeum* pages 10–72.

Rarities of ye Royal Society. By Nehemiah Grew M.D. &c. English in ink on paper: 51 leaves, written on the rectos, with occasional notes on the versos. Probably by Edward Lhwyd F.R.S. Additional MS 15076.

CONTENTS: ff.70–121. On the versos of ff.117–19 excerpts from the Catalogue of Ole Worm's Museum at Copenhagen, [Leiden 1655]: *Musei Wormiani Historia de Artificiosis*, in the hand which wrote the slip mounted on f.122 '*Domini Edwardi Lloyd Liber E Coll Jesu Oxon civilis juris studiosus*', but different from the writing of the excerpts from Grew's book.

The Royal Society has four manuscript *Catalogues of the Repository*, begun after the publication of *Musaeum* in 1681, with entries for accessions added from time to time till about 1765. Their compilation is discussed in detail, with particular account of the scientific instruments, by A.D.C. Simpson of the Royal Scottish Museum, Edinburgh: 'Newton's Telescope and the cataloguing of the Royal Society's Repository' *Notes and Records of the Royal Society* 38 (1984), 187–214.

Catalogue A (MS 413): A numbered inventory of specimens, with additions in various hands, some of the later entries being merely rough notes. A modern typed list of chapters shows that the arrangement follows Grew's printed Catalogue. Simpson attributes it to Moses Williams about 1720.

Catalogue B (MS 414): The largest of these four catalogues, begun soon after publication of *Musaeum*, with many later entries in various hands. The earliest dated acquisition is of 18 May 1661, while the latest which I noticed are from the 1730s. A modern pencilled note on the cover indicates 'ca.1741', but Simpson calls it 'an exhaustive list' by Cromwell Mortimer, secretary of the Society 1730–52, with a new classification 1753.

The leaves were sewn in thirteen pamphlets, but many are loose; numerous slips for individual entries, mostly initialled J.W. [John Woodward?] are lightly pasted in. Modern pencilled notes on the covers of the pamphlets collate them with Grew's chapters.

Catalogue C (MS 415): An interleaving of selected pages and a few of the engraved figures from *Musaeum*; a mutilated copy of the book is still in the Society's Library. The collection of printed and manuscript leaves has a preliminary label 'A General Plan of the System of Classification', and is in four sections with modern labels which relate them to Grew's chapters. The latest dated entry seems to be of 27 February 1728, but Simpson dates it 1763, presumably from the date of the Inventory MS 417 (below).

Catalogue D (MS 416): *A Complete Catalogue of the several Donations extracted from the Journal Book.* Seventeen pamphlets, sewn at the foot of the leaves. On leaf 2, a list of 19 classes: Class 1 *MSS Books* and Class 2 *Printed Books* are lacking from the text; Classes 3–19 correspond with Grew's chapters. The last entry in Chapter 19 (Coins, Antiquities &c) is dated 1734, Feb.27. Simpson states 'up to date till 1737, continued in MS 419 to 1744'. Below the list of Classes on leaf 2 is a list of periods with one or two names appended to each, beginning 1660. Nov. 28/1664.

June 15 – Dr Massey/Dr Mortimer, and ending 1726. Oct 27/1731. July 8 – Mr Theobald.

With these four Catalogues there is a later fair-written Inventory in two parts (MS 417): [1] An Inventory of the Subjects of Natural History in the Repository of the Royal Society. Nov. 17th 1763. [marked] 415, with which MS it presumably belongs. [2] An Inventory of such Antiquities . . . as are now in the Repository . . . Nov.21st 1765.

Manuscript of the Lectures on Stomachs and Guts

The Comparative Anatomy of Animals Begun. By Dr N. Grew. Feb. 8. 1676 [i.e. 1676/7]. Royal Society: RBC 5, pages 37–53.

Begins p.37^a: 'The Discourse I am now about to make, I crave leave to entitle The Comparative Anatomy of Animals Begun'. . . .; pp.37–39 [a survey of earlier comparative anatomy, defining the lecturer's present intention of 'comparison between the several parts and to compare these parts with the actions and other properties': i.e. descriptive anatomy, and physiology of function.]; p.39 'I will begin with Stomachs and Intestines': p.40–53 [dissection of mammals.]

The manuscript discourse corresponds with the printed Chapters 1–4 in *Musaeum*, second pagination 1–19, but there the three introductory pages of the manuscript were reduced to two short paragraphs.

After this first discourse Grew was requested 'to leave in the Repository those parts he should from time to time produce upon the occasion of his lectures', Royal Society, *Journal* 4, 8 Feb. 1676/7, and again on 10 July 1679 'Dr Grew shewed some draughts of the guts of some creatures which he had formerly dissected, of which he promised to bring in an account, but he did not leave the draughts'.

Printed Editions

1   *Prospectus*   1680

A folded sheet, forming two leaves printed on the first three sides; 292 × 194 mm.

collation: 1^a: Whereas a Book entituled *Musaeum Regalis Societatis* . . . [11 paragraphs promising 'impression' by Michaelmas . . .] Proposed at a Meeting of the Royal Society, February 16th 1679/80; 1^b-2^a: A List of those Persons, who have already Subscribed ... [144 names, in two columns divided on each page by a vertical rule]; 2^b blank.

Wing G1962
COPIES: British Library, Bagford's Collection (former Harleian MS 5946) ff.162–63; Bodleian Library, Oxford, Wood 658, no.794 (Russell 332). Each copy has a few contemporary but not identical manuscript corrections. I list the Subscribers with some known from other sources in Appendix I.

2   *Musaeum Regalis Societatis*   1681

[In a double rule] Musaeum Regalis Societatis. Or a Catalogue & Description Of the Natural and Artificial Rarities Belonging to the Royal Society And preserved at Gresham College. [rule] Made By Nehemjah Grew M.D. Fellow of the *Royal Society*, and of the Colledge of Physitians. [rule] Wherunto is Subjoyned the Comparative Anatomy of Stomachs and Guts. [rule] By the same Author. [rule] London, Printed by W. Rawlins, for the Author, 1681.

COLLATION: fo: π^1, A^b,B–Ddd^4, A–E^4, F^3, Tab. 1–31; engraved frontispiece, 6 ,leaves, 386 pages, 1 leaf, 1 blank leaf, 1 leaf, 43 pages, 1 blank leaf, 31 plates. Page height 327 mm.

CONTENTS:$\pi 1^b$ Engraved portrait – *Daniel Colwal Armiger Musaei Regalis Societatis Fundator*; [A]1^a title-page; [A]2^a Dedication 'To the Most Illustrious the Royal Society'; A3^{a-b} Epistle Dedicatory 'To his Honoured Friend Daniel Colwall Esq; Fellow of the Royal Society'.; A4^a–5^a The Preface; A5^b–6^a 'A Prospect of the whole Work'; A6^b [The Royal Society's Council orders] July 18th 1678 'to make a Catalogue', July 5th 1679 'That ... *Musaeum* ... be printed', Errata.; Ba^a–Ddd1^a, pages 1–386, text: 'A Description of the Rarities Belonging to the Royal Society ...'; Ddd2^a 'An Index of some Medicines'; Ddd2^b 'A List of those who have Contributed to this Musauem: excepting some Names which are lost.'; Ddd3 blank; Ddd4^a half-title: [In a double-rule] The Comparative Anatomy of Stomachs and Guts Begun. [rule] Being several Lectures Read before the Royal Society. In the year 1676. [rule] By Nehemjah Grew M.D. Fellow of the Royal Society, and of the Colledge of Physitians. [rule] London, Printed by W. Rawlins, for the

Author, 1681; Ddd4^b An Advertisement to the Reader.; A1^a–F1^a, pages 1–42, text: Of Stomachs and Guts; F2^a 'Some Notes upon the Tables'; F3 blank; Tab. 1–31.

At page 362 there is a small woodcut showing the design of Wilkins' flat floor. 'Stomachs and Guts' is an integral part of the volume, but has sometimes been catalogued in error as a separate publication. I list the Contributors (from leaf Ddd2, with a few others from the text) in Appendix I.

Headings in the Text, which vary slightly from the Contents List of 'prospect of the whole Work' printed on leaves A 5–6:

PART I   OF ANIMALS
  SECT. I. Of Humane Rarities. pp.1–10.
  SECT. II. Of Quadrupede's
    CHAP. I. Of Viviparous Quadruped's, particularly such as are Multifidous. pp.10–20.
        II. OF VIVIPAROUS QUADRUPED'S, particularly such as are BIFIDOUS and SOLIDIPEDOUS. pp.21–33.
    APPENDIX. Of certain BALLS found in the Stomachs of divers Beasts. pp.34–35.
    CHAP. III. OF OVIPAROUS QUADRUPED'S. pp.35–48.
  SECT. III. OF SERPENTS. pp.48–52.
  SECT. IV. OF BIRDS
    CHAP. I. Of Land-Fowls and of their Parts. pp.53–63.
        II. OF WATER-FOWLS, particularly of the Cloven-Footed. pp.63–67.
        III. OF PALMIPEDE'S, or WEBFOOTED FOWLES. pp.65–75.
        IV. Of the EGGS and NESTS of BIRDS. pp.75–81.
  SECT. V. OF FISHES
    CHAP. I. OF VIVIPEROUS FISHES. pp.81–103.
        II. OF OVIPEROUS FISHES, particularly such as are NOT-SCALED. pp.103–112.
        III. OF SCALED FISHES. pp.112–119.
        IV. OF EXANGUIOUS FISHES. pp.119–124.
  SECT. VI. OF SHELLS
    CHAP. I. Of whirled and single SHELLS. pp.124–141.
        II. OF SHELLS Double and Multiple. pp.141–153.
  SECT. VII. OF INSECTS
    CHAP. I. Of Insects with Naked-Wings. pp.154–160.
        II. Of Insects with sheathed-Wings. pp.160–172.

III. OF CREEPING INSECTS. pp.173–178.

PART II OF PLANTS

SECT. I. OF TREES

CHAP. I. Of WOODS, BRANCHES, and LEAVES. pp.179–186.

II. OF FRUITS; particularly such as are of the Apple, Pear and Plum-kinds. pp.186–192.

III. Of CALIBASHES, and some other like Fruits. pp.193–196.

IV. Of NUTS, and Divers other like Fruits. pp.196–211.

V. Of BERRYS, CONES, LOBES, and some other Parts of Trees. pp.211–215.

SECT. II. Of SHRUBS and ARBORESCENT plants

CHAP. I. Of SHRUBS, chiefly. pp.216–219.

II. Of ARBORESCENT Plants. pp.220–226.

SECT. III. OF HERBS

CHAP. I. Of STALKS and ROOTS. pp.226–229.

II. Of FRUITS. pp.229–232.

III. Of SEEDS. pp.232–235.

SECT IV. Of MOSSES, MUSHROOMS &c. Together with some Appendents to Plants. pp.235–241.

SECT. V. Of SEA-PLANTS

CHAP. I. Of SHRUBS. pp.242–247.

I. Of other SEA-PLANTS, and of SPONGES. pp.247–252.

PART III. OF MINERALS

SECT. I. OF STONES

CHAP. I. OF ANIMAL BODIES PETRIFY'D; and such like. pp.253–265.

II Of VEGETABLE BODIES petrify'd, and other like STONES. pp.265–274.

III. Of CORALS, and other like MARINE Productions pp.275–281.

IV. Of GEMS. pp.281–294.

V. Of REGULAR STONES. pp.294–314.

VI. Of STONES IRREGULAR. pp.315–321.

SECT. II. OF METALS

CHAP. I. Of GOLD, SILVER, and COPPER. pp.322–327.

II. Of TIN, LEAD, and IRON. pp.328–333.

III. Of ANTIMONY, MERCURY, and other METALLICK BODIES. pp.334–338.

SECT. III. Of Mineral Principles. pp.338–339.

CHAP. I. Of SALTS. pp.339–343.

II. Of SULPHURS. pp.343–346.

III. Of EARTHS. pp.346–350.

PART IV. OF ARTIFICIAL MATTERS

SECT. I. Of Things relating to CHYMISTRY, and to other Parts of NATURAL PHILOSOPHY. pp.351–357.

Of Instruments relating to Natural Philosophy. pp.357–360.

SECT. II. Of Things relating to the MATHEMATICKS; and some MECHANICKS. pp.360–368.

SECT. III. Chiefly of MECHANICKS Relating to Trade. pp.368–379.

SECT. IV. Of COYNS, and other Matters relating to Antiquity. pp.380–381.

Of COYNS. pp.381–384.

APPENDIX. Of some Particulars lately given by Dr Christopher Merret. pp.385–386.

An INDEX of some MEDICINES. p.387.

A List of those who have Contributed to this Museum: excepting some Names which are lost. p.388.

OF STOMACHS AND GUTS

CHAP. I. Of the Stomachs and Guts of Six Carnivorous Quadrupeds; sc. The Weesle, Fitchet, Polecat, Cat, Dog and Fox. pp.1–6.

II. Of the Stomach and Guts of the Mole; which seems to feed on Insects. As also of the Urchan, Squirel, and Rat; which are chiefly Frugivorous. pp.7–10.

III. Of the Stomach and Guts of such Animals as are both Frugivorous and Graminivorous; as the Rabbit, Horse, and Pig. pp.10–16.

IV. Of GRAMINIVOROUS QUADRUPEDS; a Sheep and a Calf. pp.16–19.

V. Of the Uses of the Gulet and Stomachs of Quadrupeds. And first of the Gulet. pp.20–23.

VI. Of the Uses of the Stomachs of Quadrupeds. pp.23–27.

VII. Of the Uses of the Guts of Quadrupeds. pp.27–31.

VIII. Of the Stomachs and Guts of BIRDS. pp.31–39.

IX. Of the Uses of these Parts. pp.40–42.

X. Of the Stomachs and Guts of FISHES. p.42.

Some NOTES upon the TABLES. p.43.

FINIS

NOTE: *Stomachs and Guts* I–IV: Fitchet, now a generic term for Mustelidae, here meant Stoat, Urchan is Hedgehog; VIII: *Birds described* – Cassowary, Grey owl, Cuckoo (not illustrated) Dunghill Cock, Tame pigeon, Jackdaw, Starling, Yellowhammer, Bullfinch, Wryneck, a Bunting, a Reed Sparrow much like a Bunting, House swallow; 'with notes upon others': Ostrich, Wild duck, Teal, Goose, Turkey, Japan Peacock, Duck, Carrier pigeon, Cropper Dove, Robin Redbreast, Twite, Redstart, Titlark (Meadow pipit), Water-wagtail, Solitary Sparrow (Dunnock?), House Sparrow, Chaffinch, Greenfinch, Jay. *Fish*: Jack (Pike), Tench, Barbel, Bream, Plaice, Perch, Rochet, Trout, Whiting, Cod.

All copies have the following manuscript corrections, from the printer's office, which are not in the list of Errata:

Page 62: the Maygians *crossed through*; Embroyderers *written above.*
    81: 'BACK-' *altered to* 'NECK-'
        "'Tis only' *altered to* 'Less than'
   177: 'Hirundo marinus' *altered to* 'Hirudo marina'
   181: 'Corinthian' *altered to* 'Dorick'
   239: "The AROMATICK TUBER' *crossed through*; 'A Negro Glyster Bag' *added in margin.* 'solid' *and* 'and ponderous' *crossed through.*
        'See p.385' *added at end of line 9 from foot.*
   312: 'white and pink' *crossed through*; 'all of a browne' *written above, and the* 'd' *of* 'colour'd' *obliterated.*
   343: 'Thomas' *written in space before* 'Henshaw'.
   353: Space (unfilled) for Dr Robert Witty's Christian name.

Uncorrected misprints besides those in the list of Errata include page 55 Emen [Emeu]; 136 Whired [Whirled]; 147 Cokle [Cockle]; 174 GRFAT [GREAT]; 329 Christephor Menet [Christopher Merret]; 356 hugh [hue]; 361 Brouncher [Brouncker]; 362 1699 [1669]; 372 & 376 dey [dye]; 377 wrighting [writing]; at page 362 the visit to Oxford by Cosimo III, Grand Duke of Tuscany in 1669 (Evelyn's *Diary* 16 May) is dated 1699; he came also to the Royal Society.

In many copies pages 247 and 344 are numbered 243 and 334 respectively, but some copies have one or both of these correctly paginated.

The engravings, voluntarily paid for by Colwall, are all unsigned but undoubtedly from Grew's own drawings. The portrait of Colwall, half-length to right in an oval on a blank pedestal, with the misspelt lettering below, is crudely drawn; plate-mark 263 × 160 mm. Plates 1–22 illustrate the Catalogue, plates 23–31 the Lectures; plate-mark 268 × 166 mm, except the folded plate 4 (Skeleton of a Crocodile or ye Leviathan) 163 × 520 mm. Plates 1, 2, 3, 5, 7, 8, 15, 18, 23, 24, 25, 28, 29 carry appropriate scales of inches, plates 26 and 27 have two inch-scales each; plate 4 has a scale of feet. The individual figures on each plate are lettered and partly numbered.

Wing G1952; not in Term Catalogue.

COPIES: London: British Library (2 copies) L35/87, and Eve.b.46 Evelyn's copy, see below; British Museum (Natural History) (2 copies, Grew's portrait from *Cosmologia Sacra* bound into one); Linnean Society – Linnaeus's copy, inscribed *E. Bibl. Linn. 1784 J.E. Smith*, see note at *Anatomy of Vegetables Begun* 1672 (IIA1); Royal College of Physicians, bequest of Richard Hale F.R.C.P. (1670–1728) with his armorial bookplate; Royal College of Surgeons of England, inscribed *Presented to his honoured Friend William Wagstaffe Esq from ye Author* (Grew's contemporary at Pembroke College became a London barrister); Royal Pharmaceutical Society; Royal Society of Medicine (pagination correct throughout); Royal Society (2 copies, one copy is imperfect, possibly cut up for MS 415 described above, and inscribed in a contemporary hand *Edw Harrys*).

Cambridge: Cambridge University Library – Bishop John Moore's copy for which he subscribed (M.14.22 with *Munificentia Regia 1715* bookplate), and Geoffrey Keynes copy *Bibliotheca Bibliographici 2466*; Magdalene College, Samuel Pepys's copy (*Catalogue of Printed Books 1978, no.2400*); Trinity College (see note on Newton's copy below).

Eton College – gift in 1733 of Thomas Evans, Fellow and Benefactor of the College. Oxford, Bodleian Library (4 copies): o.1.10.Med., Sherard 652 inscribed *Liber Jacobi Bobart 1681* (Bobart the Younger, keeper of the Physic Garden 1680 and first Professor of Botany on Sherard's foundation 1683), Francis Douce's (G.subt.3) and Richard Gough's (Lond.132) copies have each their armorial bookplate. Reading University, Cole 1969, no,784 inscribed *Presented to his much honoured Friend Sir John Lawrence Knight & Alderman of London from the Author* (reproduced: Cole 1944, fig.105).

University Libraries of Leeds (2), Manchester, Dublin (3), Edinburgh, Glasgow (William Hunter's Library), Amsterdam and Leiden, the Royal Colleges of Physicians of Edinburgh and of Ireland, the National Library of Scotland, and the Bibliothèque Nationale, Paris, etc.

North America (*NUC* – NR 0485181 – i.e. under Royal Society) records 42 copies including: McGill University, Montreal, *Bibliotheca Osleriana* 2840; Institute of the History of Medicine, Baltimore – 2 copies, one from the Warrington Dispensary, see Sir William Osler's note on the sale of this Library in London in 1906 (*Bibl. Osleriana* 986); Library Company of Philadelphia – 4 copies, one inscribed *Z. Isham 1681*.

Evelyn's copy (British Library, bought in 1977) has many passages marked with a pencilled line, and on the end paste-down notes of pages related to his own writings. At the description of the 'Tables of Veins and Arteries' which he gave to the Royal Society (*Musaeum* p.4) he noted 'And cost me as I remember about 20£ . . . King Charles the 2d saw them at my house at Deptford' [30 April 1663]; the price is not in the *Diary*, where he recorded buying them at Padua in 1643.

Cambridge, Trinity College – Q.11.67 has been in the Library since about the time of publication, but is not the one which Isaac Newton gave in 1680–[81]. (Personal study of the extant copy, and fuller information kindly given me by David McKitterick, the Librarian). Newton's own copy is in the Babson Institute, Wellesley, Massachusetts (Newton Collection no.406), see J.R. Harrison *The Library of Isaac Newton*, 2nd ed. 1978, no.716. Harrison records John Locke's copy in *The Library of John Locke* 1965 no.1321.

The original sheets were reissued with cancel title-pages, by Thomas Malthus 1685 and 1686, Samuel Holford 1686, and Hugh Newman 1694:

3   *Thomas Malthus*   1685

[In a double rule:] Musaeum Regalis Societatis: or, A Catalogue and Description of the Natural and Artificial Rarities Belonging to the Royal Society, and preserved at Gresham Colledge. [rule] Made by Nehemiah Grew, M.D. Fellow of the Royal Society, and of the Colledge of Physitians. [rule] Whereunto is Subjoyned the Comparative Anatomy of Stomachs and Guts. [rule] By the same Author. [rule] London, Printed for Tho. Malthus, at the Sun in the Poultrey, 1685.

COLLATION: The sheets of 1681 reissued with cancel portrait and cancel title-page; the half-title for 'Stomachs and Guts' is the original of 1681; the publisher's manuscript corrections are all present.

R. White delin.   t. Sculp: 1683.

DANIEL COLWAL Armiger.
Musæi Regalis Societatis Fundator.

PLATE 7   *Daniel Colwall* F.R.S.
Portrait by Robert White, from *Musaeum* 1685

PORTRAIT: Daniel Colwall by Robert White, line engraving, half-length to right in oval, architectural surround with blank cartouche; lettered below the surround, at left *R. White delin,* at right *et Sculp 1681,* and at foot *DANIEL COLWAL Armiger/Musaei Regalis Societatis Fundator.* This portrait, though dated 1681, was first published here in 1685; it is in every way better than the portrait in the edition of 1681; plate mark 20 × 16.5 cm, engraved surface 17 × 15 cm. [Plate 7.]

Wing G1952A; Russell 334; *NUC*-NR 0485183. Not in Term Catalogue.

COPIES: National Library of Scotland; Harvard University – Zoology Library; Princeton University; W.A. Clark Library, Los Angeles.

4   *Thomas Malthus*   1686

As the previous issue, but the imprint is dated 1686.

Wing G1954A.

COPIES: Royal Society; Leeds University, Brotherton Library; Utrecht University Library.

NOTE: The Royal Society's copy carries the bookplate of 'The Right Honble Charles Earle of Winchelsea Viscount Maidstone Baron Fitz-Herbert of Eastwell 1704' [Charles Finch, 4th Earl 1672–1712]. Page 247 is correctly numbered in the Leeds copy but as '243' in the Royal Society copy; in both copies page 344 is misnumbered '334'; the Leeds copy lacks the portrait.

5   *Samuel Holford*   1686

As Malthus's reissue, but with imprint: LONDON, Printed for S. Holford, at the Crown in the Pall-Mall, 1686.

COLLATION, including Portrait, as in Malthus 1685.

Term Catalogue, Michaelmas [November] 1685 (TC II, 150) – Holford's imprint without date; Wing G1954. Wing (first edition) G1953 'S. Holford 1685, National Library of Scotland' was a ghost entry.

COPIES: Wellcome Institute, London; Cambridge University Library – Lib.3.68.3: *Munificentia Regia 1715* bookplate, Bishop John Moore's copy, evidently alienated and regained (printed label 'The gift of Spencer George Perceval... 1921' covering his ownership label), with ownership signature of Edward Butt of Wimborne and label of T. Bell Salter M.D.; Edinburgh and St Andrews University Libraries. United

States (*NUC–NR* '82): Harvard and Columbia Universities.

6 *Hugh Newman* 1694

Musaeum . . . By the same Author. [as 1681]. Illustrated with a great number of Cuts curiously Engraven on Copper Plates. London, printed for Hugh Newman at the Grashopper in the Poultrey, 1694.

COLLATION: The original sheets with a cancel title-page; the publisher's manuscript corrections are all present; page numbers at p.247 and 344 vary among copies as in 1681. Portrait as in 1685.

Term Catalogue (TC II.530) Michaelmas [November] 1694; 'Reprinted'. Wing G1955; Russell 336.
COPIES: British Library 38.f.14; British Museum (Natural History); Royal College of Obstetricians & Gynaecologists (cancel and cancellans title-pages); Royal Society (cancel and cancellans title-pages); Wellcome Institute, London. Reading University Library, Cole 1969 no.786. Dublin: Trinity College. Edinburgh: Royal College of Physicians – 2 copies; University Library. Glasgow: Royal Faculty of Physicians and Surgeons; University Library. Wellington, New Zealand – Turnbull Library. United States (*NUC–NR* '84): National Library of Medicine, and Smithsonian Institution, Washington; New York – Botanical Garden, and Public Library; Yale University Library.

IV   SEA-WATER MADE FRESH

Chronology

The precise sequence of editions (A1–D4) of Fitzgerald's and Grew's pamphlets cannot be determined, since some are undated and others carry only a year date.

1674   Robert Boyle. *The Saltness of the Sea* London 1674 (Fulton 113).

1675   William Walcot's Patent, 28 October 1675.

1678   William Walcot's Bill considered by Lords Committee 13 March 1677/8.

1679   Grew's Royal Society discourses on London waters, 5 June and 3 July 1679.

1683   Robert Fitzgerald's Patent, 9 June 1683.

Fitzgerald. *Salt-water sweetned*, including Boyle's *Letter to Beale*. London 1683. B1.

Fitzgerald: [*The same.*] Dublin 1683. B2.

[Grew]. *Sea-water made fresh*. London 1683. A1.

Fitzgerald. Latin translation: *Agua salsa dulcorata*. London 1683. B16.

Fitzgerald. French translation: *L'Eau de mer douce*. [London] 1683. B12.

Fitzgerald. Spanish translation: *Agua salsada dulce* [London 1683 or '84?]. B19.

Fitzgerald. *Salt-water sweetned*, 2nd edition. London 1683. B3.

Grew. *Sea-water made fresh*, 2nd edition [not traced]. A2.

Fitzgerald. *Salt-water sweetned*, 3rd edition. London 1683. B4.

Grew. *Sea-water made fresh*. 3rd edition [not traced]. A3.

Boyle 'Appendix to his Letter', *Philosophical Transactions* 30 October 1683.

Order in Council vacating Walcot's Patent, 31 October 1683.

1684   Grew. Latin translation: *Circa aquam marinam dulcoratam*. London, 14 February 1684. A13.

Fitzgerald & Grew. [English pamphlets together] 4th edition. 14 February 1684. A4, B5.

Grew. [French translation]. Londres 1684. A11.

Fitzgerald & Grew. [French translations together]. 4me édition. Londres 1684. B15.

Fitzgerald & Grew. [English pamphlets together] 5th edition. London, 8 March 1684. A5, B6.

Fitzgerald. [Latin translation] 2nd edition. London 1684. B17.

Fitzgerald. [German translation] Hamburg 1684. B18.

Fitzgerald & Grew. [English pamphlets together] 6th and 7th editions [not traced]. A6–7, B7–8.

Fitzgerald. *Salt-water sweetned & Supplement*. 8th edition. London 23 July 1684. B9.

Grew. *Sea-water made fresh*. 8th edition. London, 22 May [and reissue 12th August] 1684. A8.

[Agreements &] *Conditions*. London 1684; issued also in Fitzgerald's and in Grew's 8th editions, May, July, August 1684, A8, B9, C1–3.

Grew. *Sea-water made fresh*. 9th edition. London, 29 September 1684. A9.

Fitzgerald. *Salt-water sweetned. Further additions*. 9th edition, London, 13 October 1684. B10.

Fitzgerald & Grew. *A Brief of Two Treatises*. Broadsheet [London 1684?]. C4.

Fitzgerald & Grew. [A Brief... Italian translation] *Estratto di due Trattati* [London 1684?]. C6.

1685   Fitzgerald & Grew. *A Brief*... 2nd issue [after 6 February 1685]. C5

Grew. *Sea-water made fresh*. 10th edition. London, 25th March 1685. A10.

Fitzgerald. *Salt-water sweetned*. 10th edition. London, 3 April 1685. B11.

[Kempthorne]. *Certificates of several Captains*... London, 1685. C7.

1693   Walcot's Bill before Lords Committee 28 February 1692/3; Fitzgerald given leave to be heard on his Petition 7 March; his Counsel heard 10 March, Fitzgerald offered Proposal, the Bill was negatived.

1694   Walcot's Bill again considered 21 March 1693/4.

1695   Walcot's amended draft of a Salt Water Act before Lords Committee 21 February 1694/5; Walcot attended and Fitzgerald's Letter read 13 March 1694/5.

Walcot. *An answer to Mr Fitzgerald's State of the Case*. London 1695. C8, D1–2.

1697   *A Treatise concerning the Exercise of William Walcot*. London 1697. D3.

1702   H. Walcot. *Sea-water made fresh and wholesome*. London 1702. D4.

IVA   SEA–WATER MADE FRESH   1683

1   New Experiments, And useful Observations concerning Sea-Water, made Fresh, according to the Patentees Invention: in a Discourse humbly dedicated to His Majesty the King of Great Britain, &c. [rule] By a Fellow of the Colledge of Physicians, and of the Royal Society. [rule] Printed Anno Dom. 1683.

COLLATION: 8^0: A–C^8, π^1; 2 leaves, 44 pages, 1 blank leaf. Page height 148 mm.

CONTENTS: A^1 blank, A^{2a} title-page, pages 1–44 text; a blank leaf. Page-numbers in round brackets at the centre of the head margin.

Grew's undated *Memorial to the King* (Sloane MS 2145, ff.270–71) appears to be a draft for the Dedication of this pamphlet.

Wing G1956.
COPIES: British Library; Wellcome Institute, London; Cambridge University Library; Bodleian Library, Oxford.

The same: another issue: Printed by John Harefinch 1684.

Wing G1956A.
COPIES: London, Guildhall Library; Harvard University.

2–3   *Second and Third Editions:* No copies traced

4   *Fourth Edition:* 14 February 1684

New Experiments... Royal Society. [as First edition] [rule] The Fourth Edition. February the 14th [rule] London, Printed by John Harefinch, Anno Dom. 1864.

COLLATION: sm. 4^0: A–B^4; 1 leaf, 13 pages. Page height 195 mm.

CONTENTS: A^{1a} title-page, A^{2a}–B^{4a} text.

Wing G1957, *NUC*–NG '70.
COPIES: British Library; Cambridge University Library; Bodleian Library and Christ Church, Oxford; Harvard; Yale.

NOTE: A few words changed from first edition and misprints corrected.

5   *Fifth Edition:* 8 March 1684

New Experiments... Royal Society. [as First edition] [rule] The fifth Edition. March the 8th [rule] London, Printed by John Harefinch in Mountague-Court in Little Britain. 1684.

COLLATION: 3 leaves: 1^a Title-page, 2^a–3^b text, the end wanting. Page height 132 mm.

Wing G1957A.
COPY: Library Company of Philadelphia (E. Wolf 2nd, *Check-list supplementary to Wing,* 1959).

6–7   *Sixth and Seventh Editions,* 1684: no copies traced

8   *Eighth edition, 2 issues* 1684

New Experiments... Royal Society [as First edition] [rule] By Nehemiah Grew M.D. Fellow... [rule] The Eighth Edition. May 22nd [2nd issue August 12th] [rule] London, Printed by John Harefinch in Mountague-Court in Little Britain. 1684

COLLATION: 8°: A^8B^4; 1 leaf, 20 pages, 1 blank leaf. Page height 145 mm.

Wing G1957B and 1958, *NUC–NG* '71 and '72.
COPIES: 1st issue Yale; 2nd issue British Library 712.g.17(13) with Fitzgerald's 'Agreements and Conditions...' bound between pages 4 and 5; Bibliothèque Nationale, Paris; Library of Congress; Columbia University, New York; Harvard.
British Library 712.g.17 is a collection of 23 pamphlets including several of Fitzgerald's, Grew's and Walcot's 'Salt-water' papers.

9   *Ninth Edition*   29 September 1684

New Experiments... Royal Society. [as Eighth edition] [rule] The Ninth Edition. Septemb. 29. [rule] London. Printed by John Harefinch in Montague-Court in Little Britain. 1684.

COLLATION: 8°: B^{7-8}, C^8, D^1; 1 leaf, 20 pages. Page height 145 mm.

Wing G1958 A & B, *NUC*-NG '73.
COPIES: British Library – 3 copies 117.a.16(2),712.g.17 (17 & 19); Bod-

leian Library, Oxford – Ashmole C. 67(7); Columbia University, New York.

Though dated earlier this was issued as the second part (Wing G1958B) of Fitzgerald's *Farther Additions*, 9th edition, 13 October 1684 (B10) as the signatures indicate.

10   *Tenth Edition*   25 March 1685

New Experiments... Royal Society. [as Eighth edition [rule] The Tenth Edition, March 25. [rule] London, Printed by John Harefinch in Mountague-Court in Little Britain. 1685.

COLLATION: 8^0: C^8–D^4; 1 leaf, 20 pages, 1 blank leaf. Page height 15 cm.

Wing G1958C, *NUC–NG* '74.
COPIES: British Library T 1834(2); Cambridge University Library – Geoffrey Keynes copy *Bibliotheca Bibliographici* 2467; New York Public Library.

NOTE: Issued with Fitzgerald's *Farther Additions*... 10th edition, April 3, 1685, followed by *The Agreements and Conditions*..., and *Certificates*.

11   *New Experiments:* French translation 1684

Nouvelles Expériences et utiles remarques, sur l'Eau de Mer Dessalée, suivant la Nouvelle Invention du Sieur Fitzgerald, & ses Associez. Comprises Dans ce Traité, tres humblement Dedié à Sa Majesté Britannique Par Un Membre du Collège des Médecins de Londres, & de la Société Royale. [rule] Et Traduit de l'Anglois par le Sieur Guy Miege. [rule] A Londres, 1684.

COLLATION: 4^0: A^4, B^3; 1 leaf, 12 pages. Page height 18 cm.

Wing G1958D.
COPY: British Library 712.g.17 (22).

12   Another issue in Fitzgerald *Additions au Traité*... 1684 (B15).

13   *New Experiments:* Latin translation [1684?]

Dissertatio quae complectitur Nova Experimenta utilesque Obervationes circa aquam marinam, secundum eorum Artem, quibus

Regium Diploma concessum est, Dulcoratam. Serenissimo Principi Carolo II. Magnae Brittanniae Regi Humillime oblata. [rule] Offerebat Nehemias Grew, M.D. & Societatis Regiae Londinensis utriusque Socius. Londini, Excudebat J. Gain.

COLLATION: 8^0: A^8, B^7; 29 pages. Page height 16 cm.

CONTENTS: pp.1–17 Three Propositions; pp.17–22 Boyle's letter to Beale; pp.23–24 Letters Patent, 2 November '1612' [1683]; pp.28–29 Letters Patent 28 February 1683/4.

Wing G1949, *NUC*–NG '64.

COPIES: British Library 712.g.17 (11); Cambridge University Library, Ddd.25.229.5; Bodleian Library, Oxford – Ashmole D 60(3), pp.1–24 only, following Fitzgerald's second Latin edition (B17); Christ Church, Oxford; Bibliothèque Nationale, Paris; University of Wisconsin, Madison.

IVB   FITZGERALD'S PAMPHLETS

1   Salt-water Sweetned ... [includes] A Letter of Mr Boyle to the learned Dr John Beale [not previously printed, Fulton 236–37] and Approbation signed by several Fellows of the Royal College of Physicians. London, Printed for Will. Cademan 1683.

COLLATION: 4^0: 1 leaf, 18 pages. Page height 188 mm. Dedication to the King signed 'R. Fitzgerald'. Boyle in his paper in *Phil. Trans* 30 October 1683 refers to this pamphlet as 'lately published'.
Wing F1087.

COPIES: British Library 712.g.17(7); Cambridge University Library Ggg.138.8; Oxford, Bodleian Library – 2 copies. United States (*NUC*–NF 0172022): National Library of Medicine; Harvard; Yale; Huntington and W.A. Clark Libraries, Los Angeles.

2   —— Another edition: Dublin, John Ray 1683. 4^0.

Dix *Dublin-printed books 1601–1700*, 200, Wing F1088.
COPIES: Trinity College Dublin; Cashel Diocesan Library.

3   —— Second edition. London, for Will. Cademan 1683. 4^0.

Wing F1087A.
COPIES: Cambridge: University Library M.17.26.2, and St John's College Library.

4 —— Third edition. London, for Will Cademan 1683 4⁰.

Wing F1088A.
COPY: Harvard.

5 —— Fourth edition. 14 February. London by John Harefinch 1684. 4⁰: 1 leaf, 13 pages.

Wing F1089; *NUC–NF* '23.
COPIES: British Library 117.c.28; Cambridge University Library Bbˣ.10.39(7) (E); Oxford, Bodleian Library G Pamph. 1120(19) before Grew's edition of same date [A4]; New York Academy of Medicine; University of Wisconsin, Madison.

6 —— Fifth edition. London by John Harefinch 1684. 8⁰.

Wing F1089A
COPY: Library Company of Philadelphia, with A5 above.

7–8 —— Sixth and Seventh editions; no copies traced.

9 —— The Supplement to . . . Salt-Water sweetned. 8th edition, July 23. London by John Harefinch [1684], 8⁰: 21 pages. *Includes* Conditions . . . for the Use of this Invention.

Wing F1090, *NUC–NF*, 24.
COPIES: British Library 712.g.17 (12) and (15); Bibliothèque Nationale, Paris; Library of Congress, Washington; Yale University Library.

10 —— Farther Additions to . . . Salt Water sweetned. 9th edition. Octob. 13. London, by John Harefinch 1684.

COLLATION: 8⁰: A⁸, B⁶; 3 leaves, 21 pages. Height 145 mm. Issued with Grew's 9th edition (dated Septemb. 29) which has continuous signatures [A9].

Wing F1082, *NUF–NF* '18

COPIES: British Library, 3 copies – 117.a.16(1), 712.g.17 (16 and 18); Oxford, Bodleian Library – Ashm.C.67(6); New York, Columbia University Library.

11 —— Farther Additions... The Tenth Edition. April 3. London, by John Harefinch 1685.

COLLATION: 8^0: A^8, B^4; 13 leaves, 22 pages. Page height 14 cm.

CONTENTS: pp.11–16 Boyle's Letter, p.18 Macdonell's Letter; pp.21–22 the King's Order for publication. Issued with Grew's 10th edition, dated March 25, [A10] and followed by 'The Agreements and Conditions' (E$^{1-3}$) and 'Certificates' [C2 and C7] (F$^{1-2}$); E^4 is blank.

Wing F1084; *NUC*–NF '20
COPIES: British Library T 1834(1); Cambridge University Library, Geoffrey Keynes copy *Bibliotheca Bibliographici* 2119; New York Public Library.

12 *French*: L'Eau de mer douce... [London] 1683.

Wing F1081.
COPY: Bibliothèque Nationale, Paris.

13 —— Second edition not traced.

14 —— Third edition [London] 1684

Wing F1081A
COPY: Manchester University Library.

15 —— Fourth edition: Additions au Traité... traduit par le Sieur Guy Miege. Londres 1684.

COLLATION: 8^0: A–B^8, 3 leaves, pp.1–25 [with] C^8, D^4, 1 leaf, pp.29–50 Nouvelles Expériences.. sur l'Eau de Mer dessalée... par Nehemie Grew. [A12].

Wing F1077.
COPY: British Library 712.g.17 (21).

16 *Latin:* Aqua salsa dulcorata... Londini, Impensis Edvardi Brewster ad Insigne Pelicani in Coemeterio Paulino 1683.

COLLATION: 4^0: A–D^4, 32 pages. Page height 188 mm.

Wing F1078; *NUC*–NF '17.
COPIES: British Library 712.g.17(9); Cambridge University Library M.17.26(1); Oxford, Christ Church; W.A. Clark Library, Los Angeles; Yale.

NOTE: The list of Physicians signing the Approbation [see B1] omits Drs Weatherby and Andrew Creagh.

17 —— Second edition. Aqua salsa dulcorata . . . Editio secunda. Excudebat J. Gain pro Gulielmo Cadman. 1684.

COLLATION: 8^0: A^8, B^4; 24 pages. Page height 16 cm.

Wing F1079.
COPIES: British Library 712.g.17(10); Cambridge University Library Ddd.25.229 (4); Oxford: Bodleian Library – Ashmole D 60 (2) before Grew's Latin edition [A13], and Christ Church.

18 *German*: Das Süsse-gemachte, Saltz-Wasser oder Eine genaue Beschreibung des neues Kunststücks und endlich ein Schreiben des geehrtes R. Boyle. Hamburg 1684. 12^0.

COPY: Oxford, Bodleian Library – Vet.D.3.f.65.

19 *Spanish*: Hasser de Agua Salsada Dulce . . . [London 1683 or '84]

COLLATION: 4^0: A^4, 8 pages. Page height 225 mm.

Wing F1085.
COPY: British Library 712.m.1(23)

IVc   FITZGERALD'S SUPPLEMENTARY PAMPHLETS

1   The Conditions upon which the Patentees for making Salt or Brackish Water Fresh and Wholesom [*sic*] do intend to conclude

with such Persons that shall please to agree with them for the Use of this Invention either by Sea or Land. July 2, 1684. London, John Harefinch 1684.

COLLATION: Single sheet, printed on both sides, height 288 mm.

COPY: British Library 712.m.1(24), with List of 'Patentees and others concern'd: Earl of Berkeley, Viscount Falkland, R. Fitzgerald, Nat. Johnson, Theo. Oglethorpe, Tho. Maule, W. Bridgman, P. Trant, George Doddington.

2   The Agreements and Conditions upon which Robert Fitzgerald, Theo. Oglethorpe, Tho. Maule, W. Bridgman, P. Trant, Patentees... have already concluded with several Merchants Commanders and Owners of Ships, and do intend to conclude with such other Persons as are willing to agree with them for the use of the Invention, either by Sea or Land.

COLLATION: 8^0: E^4 (1–3 text, not paginated; 4 blank). Page height 15 cm.

COPY: British Library 712.g.17 (14) bound between leaves A3 and 4 of Grew: New Experiments 8th edition, 1684 [A8].

3   ——— : Another issue: 'Conditions . . .'

COLLATION: 8^0: C^{1-3}; 6 pages; 145 mm.

COPY: British Library 712.g.17 (15).

NOTE: Continuous with Grew's 8th edition, also issued with his 10th edition [A8 and 10].

4   *A Brief of Two Treatises*   [1684?]

[Caption title]: A Brief of two Treatises formerly Published concerning the making Sea-water Fresh, and of some Matters transacted since in relation to the same.

COLLATION: Single sheet printed on both sides, height 276 mm. No imprint, [London 1684?]

CONTENTS: *Recto*: [Text begins] – The first Book printed by the Patentees dedicated to his Majesty... asserts [summary of Fitzgerald's *Saltwater sweetned*]. [List of 23 physicians]... A letter written by ... Mr Boyl.... Mr Boyl's Experiment made before the King... A letter of the Third of November 1683...

Verso: The second Treatise, is written by Doctor Nehemiah Grew [summary of *Sea-water made Fresh*.]... Since the publishing the said Treatises, His Majesty has ordered the following Certificate... Viz.... the Patentees Engine... may be placed on Ship-board without Inconvenience or Danger... [signed by] Sam Chamblet, Deputy-Master of Trinity House [and 22 others]... Sir Charles Scarborough... and other Physicians have approved this prepared Water... the use thereof in Coffee, Chocolet and Tea... FINIS.

Published presumably between November 1683 and the King's death on 6 February 1684.

Wing F1080, as by Fitzgerald.
COPIES: British Library 816.m.7 (135) dated in the General Catalogue [1687?]; Oxford, Christ Church – 2 copies.

5 *A Brief*... [Second edition 1685]

[Caption title]: A Brief of two Treatises... [as previous edition].

COLLATION: Single sheet printed on both sides, height 290 mm. No imprint, [London 1685]

CONTENTS: Text as in first edition, except that each Treatise is described as 'Dedicated to his late Majesty', and the additional paragraphs include mention of letters from Barbadoes 29 January 1684/5 and 'Mevis' [Nevis] 21 February 1684/5.

Published presumably in 1685 following the final, tenth edition of the two Treatises.

Wing F1080A.
COPY: British Library 645.e.1.(46) dated in the General Catalogue [1687?].

6 *A Brief*... Italian translation [London 1684?]

[Caption title]: Estratto di due Trattati che sono comparsi doppo qualche tempo sopra il secreto trovatosi per render dolce L'acqua del Mare...

COLLATION: Single sheet printed on both sides, height 288 mm. No imprint, [London 1684?] Translated from the first edition and printed in close imitation of its style.

Wing F1081B and G1949A.
COPY: British Library 712.m.1 (22^{xxx}) dated in the General Catalogue [London 1695?].

7   Certificates   1685

The Certificates of several Captains and Masters of Ships, and others, both at Sea and Land, who have used the Patentees Engine for making Salt Water Fresh. Printed by John Harefinch in Mountague-Court in Little Britain, 1685.

COLLATION: Single sheet printed on both sides, height 293 mm.

CONTENTS: Letters from John Kempthorne, October 22, 1684 and Robert Crauford, Lieutenant Governor, Fort, Sheerness, February 2, 1685; *lower part of verso*: The opinions of several Eminent Physicians of the Colledge [signed with forty-seven names, including Grew, in four columns].

Wing C1769, this copy alone.
COPY: British Library 816.m.7 (102) under *Kempthorne*, with cross reference from Fitzgerald.

The Certificates... reprinted in Fitzgerald's *Farther Additions...* 10th ed. 1685. [B11].

8   State of the Case   1692

The State of the Case concerning the Patent of making Salt Water Fresh as it stands between Mr Walcot of the one part, and the Lord Falkland, William Bridgman Esq; Mr Fitz-Gerald, and several others on the other part, – in William Walton *An Answer to Mr Fitz-Gerald's State of the Case* 1695, leaf A2^{b}. (See 1693–95 in the preceding chronology for the discussions before the House of Lords Committee).

IVd   WALCOT'S PAMPHLETS

1   An Answer to Mr Fitz-Gerald's State of the Case concerning the Patent of Making Salt Water Fresh [rule] Licens'd, Jan. 16. 1694/5. D. Poplar [rule] London, Printed in the Year 1695.

COLLATION: Fo: A–E², F¹; 2 leaves, 22 pages. Height 29 cm.

CONTENTS: A¹ᵃ title-page; A²ᵃ 'A Preliminary...'; A²ᵇ 'Their State of the Case...'; B¹ᵃ–F¹ᵇ 'An Answer...'.

Wing W285A.
COPIES: British Library 712.m.1(22); Huntington Library.

2   The Letters Patent of the States-General (Translated out of Dutch) granted to William Walcot, Esq: Upon their Seeing his Demonstration of his Art of Making Sea-Water Fresh, &c. And Putrified Water Wholesome. [No imprint]. (Presumably issued with Walcot. An Answer... 1695).

COLLATION: Fo: A–B²; pages 1–8.

CONTENTS: pp.1–2 States General; pp.3–5 States of Holland; pp.7–8 States of Zeeland. Patents issued 1684, translated 1692.

COPY: with the foregoing, numbered (22ˣ) in B.L. copy.

3   A Treatise concerning the Exercise of William Walcot, Esq. his Art of making Salt-Water Fresh &c... London, A. Roper and B. Aylmer, 1697. F⁰: 14 pages.

COPIES: British Library 712.m.1(22ˣˣ); Cambridge University Library J.9.33. The text includes the three series of Dutch Letters Patent.

4   Sea-Water made Fresh and Wholesome [sic]. Objections against the Use thereof Removed, and the Advantages proved to be very Great. By H. Walcot. London, for R. Parker 1702. 4⁰: 4 leaves, 28 pages.

COPY: British Library 712.g.17 (23).

NOTE: The text refers to William Walcot's Patents of 1675 and 1692.

V A TREATISE OF THE BITTER PURGING SALT: SAL CATHARTICUS 1695

1 [In a double rule]: Tractatus de Salis Cathartici Amari in Aquis Ebeshamensibus, et Hujusmodi Aliis Contenti Natura & Usu. [rule]. Authore Nehemia Grew M.D. Utriusque Regiae Societatis Socio. [rule] Londini: Impensis S. Smith & B. Walford ad Insignia Principis in Coemeterio D. Pauli. 1695.

COLLATION: 12⁰: A⁶, B–E¹²; 6 leaves, 96 pages. Page height 145 mm.

CONTENTS: A1ᵇ Imprimatur of Royal Society: Mart.27.1695 and of Royal College of Physicians: Maij 3.1695; A2ᵃ Title-page; A3ᵃ–4ᵇ Dedication 'Inclytae Celeberrimaeque Utrique Regiae Societati' with running headline: Epistola Dedicatoria; A5ᵃ–6ᵇ Elenchus Capitum (with Errata, 2 lines at foot of A6ᵇ), running headline: Elenchus Capitum); pp.1–47 text; Pars Prior de Salis Cathartici Amari Natura, with running headline across openings of pp.2–3 to 46–47 De Salis Cathartici/Amari Natura: C12ᵇ blank; pp.49–96 text Pars Altera de Salis Cathartici Amari Usu, with running headline across openings of pp.50–51 to 94–95 De Salis Cathartici/Amari Usu. and on p.96 De Salis Cathartici, &c.

CHAPTER HEADINGS: see the comparative list below.

NOTE: The Royal Society's imprimatur is signed by the President, Robert Southwell, and the R.C.P's by the President, John Lawson, and the Censors: Samuel Collins, Richard Torless, Edward Tyson and Martin Lister. A printed marginal note at p.42 refers to Grew's *Luctation*: 'In posteriore Exercitatione de Mixtione'. The Errata note on leaf A6ᵇ is not in all copies. The title-page has been reproduced by A. Sakula (1984) p.11.

Wing G.1959. Term Catalogue, Trinity [June] 1695 (TC II,559).
COPIES: British Library 1171.c.12 without Errata; Royal College of Surgeons of England, without Errata; Wellcome Institute, London: Cambridge University Library z.14.21, Bishop John Moore's copy, small *Munificentia Regia 1715* bookplate on title verso; Bodleian Library, Oxford – 2 copies: Ashmole C 64, and Gough Addn Worcester 8⁰.9; Trinity College Dublin; Bibliothèque Nationale, Paris; Amsterdam University Library. United States (*NUC–NG* '79 & '80): National Lib-

rary of Medicine, inscribed on fly-leaf and title-page *Nicholas Stonestreet*; Dickinson College, Carlisle, Pennsylvania; W.A. Clark Library, Los Angeles; New York Academy of Medicine; Yale Medical Historical Library.

2 *Bitter Purging Salt*   Moult's pirated translation 1697

[In a single rule]: A Treatise of the Nature and Use of the Bitter Purging Salt Contain'd in Epsom, and such other Waters. [rule] By Nehemiah Grew, M.D. Fellow of the College of Physicians, and of the Royal Society. [rule] London, Printed in the year 1697.

COLLATION: 8°: A–D⁸; 64 pages. Page height 165 mm.

CONTENTS: A1ᵃ Title-page; A1ᵇ blank with 'Advertisement' pasted onto it; A2ᵃ⁻ᵇ 'To the Reader'; A3ᵃ–D8ᵇ, pages 5–64 text. For chapter headings see the comparative list below.

Running heading, across openings, pp.6/7 to 62/63: The Nature and Use of/the Bitter Purging Salt.

Printed slip pasted on verso of title-page: 'ADVERTISEMENT/ That this Salt is made and sold in greater or lesser Quantities, by *Francis Moult* Chymist, at the Sign of *Glaubers-Head* in Walting–Street; And this Translation at no other Place'. The slip appears to have been cut from a sheet of copies, for the Cambridge example has at its lower edge part of the heading of a second copy.

The Address 'To the Reader' is unsigned, but states ' . . . other Medicines that have grown Popular were ushered out with printed Directions . . . This I thought sufficient to Vindicate my Translation of Dr Grew's Nature & Use of the Bitter Purging Salt, which I intend to give to those who buy any quantities of the Salt . . . I doubt not his Pardon for my Translating it without his knowledge'.

Not in Term Catalogue; Wing G1960.
COPIES: British Library – 7470.aaa.25. Cambridge University Library – Bbˣ.12.11ⁱ(F), Bishop John Moore's copy with small *Munificentia Regia 1715* bookplate. John Rylands University Library, Manchester. United States (*NUC–NG* '80.1): Folger Library, Washington D.C.; Kansas University Medical Center Library; Yale Medical Historical Library, J.F Fulton's copy.

3   *Bitter Purging Salt*   Bridges's authorised translation, 1697

[In a double rule]: A Treatise of the Nature and Use of the Bitter Purging Salt. Easily known from all Counterfeits by its Bitter Taste. Written Originally in Latin, by Nehemiah Grew, Dr. in Physick, Fellow of the College of Physicians, and of the Royal Society. And now published in English, By Joseph Bridges, Dr. in Physick. With Animadversions on a late corrupt Translation publish'd by Francis Moult, Chymist. [rule] London, Printed by John Darby, for Walter Kettilby, at the Bishop's Head in St *Paul's* Church-yard, 1697.

COLLATION: 8⁰: A–F⁸, G–H²; xvi, 88 pages. Page height 155 mm.

CONTENTS: A1ᵃ Half-title; [between rules] A Treatise of the Nature and Use of the Bitter Purging Salt.; A1ᵇ Imprimatur of Royal College of Physicians, without date, but signed by Sir Thomas Millington, President [since October 1696]; A2ᵃ title-page; pp.v–viii 'To the Reader'; pp.ix–x Latin approbations from Royal Society, Mart. 27.1695 and from Royal College of Physicians, undated, with conclusion of preface; A6ᵇ–7ᵇ, pp.xii–xiv Grew's dedication to Royal Society and Royal College of Physicians; A8ᵃ⁻ᵇ. pp.xv–xvi The Contents; B1ᵃ–C7ᵃ, pp.1–29 Part the First. Of the Nature of the Bitter Purging Salt; C7ᵃ–E7ᵃ, pp.29–61 Part the Second. Of the Use of the Bitter Purging Salt; E7ᵇ–G2ᵃ, pp.62–83 A Table showing some of the most Egregious Falsifications made in the Translation by Francis Moult. G2ᵇ blank; Haᵃ–2ᵇ, pp.85–88 Postscript . . . the Narrative of Mr Josiah Peter . . . 4th August 1697.

CHAPTER HEADINGS: see the comparative list below.

Term Catalogue Michaelmas November 1697 (TC III, 40). Wing G1960A.

COPIES: Royal College of Surgeons of England – bookplate of Archdeacon Thomas Sharp (1693–1758); Royal Society of Medicine – 2 copies (1) inscribed at end of Imprimatur *August 20.1697* with a contemporary note of other waters on title-page, (2) lacks Josiah Peter's *Narrative*, ownership initials *T.S.* on title-page; Wellcome Institute, London. Oxford: Bodleian Library (Gough Nat.Hist.68), Christ Church, and Magdalen College. Edinburgh, Royal College of Physicians. United States (*NUC*–NG '80–81): Folger Library, Washington DC; Massachusetts Historical Society, Boston; University of Kansas Medical Center; Library Company of Philadelphia; Yale Medical Historical Library – J.F. Fulton's copy.

NOTE: Bridges's preface, p.vii, noted that in Moult's edition 'the Printer's name is not given, though the Author dareth to own the libel and to give notice of it in the London Gazette'. At p.xiii Grew's statement that he began this research 'more than fifteen years past' is qualified by the footnote 'Now near eighteen Years'. Moult's 'falsifications' are tabulated at length on pp.62–83, and his refusal to apologise or withdraw is described on pp.85–88. On 15 July 1698 Grew obtained a Patent to protect his monopoly of making Epsom Salt.

Chapter Headings in (G) Grew's first edition 1695, (M) Moult's piracy 1697, and (B) Bridge's translation 1697.

> *Part I*
> G = De Salis Cathartici Amari Natura.
> M = Of the Nature of the Bitter Purging Salt.
> B = [The same].

1 G. De modis quibus Aquae Catharticae Amarae primo innotuere.
   M. How the Bitter Purging Waters were first discovered.
   B. Of the Means whereby the Bitter Purging Waters came first to be generally known.

2 G. De Aquarum Catharticarum Amarescentium Natura.
   M. Of the Nature of the Bitter Purging Waters.
   B. [The same].

3 G. De Salis Aquae Catharticae Proprii, hoc est Amari, Natura.
   M. Of the Nature of the Bitter Salt, peculiar to the Purging Waters.
   B. Of the Nature of the proper, that is, of the Bitter Salt of the Purging Waters.

4 G. De Salis Cathartici Amari, ab Alumine, & Sale Muriatico seu Communi Differentia.
   M. Of the difference between the bitter Purging Salt, Allom, and the Muriatick Salt.

B.   Of the Difference of the Bitter Purging Salt, from Alum, and from common Salt.

5   G.   De Salis Cathartici Amari, a Salibus Nitrio, & Calcario Differentia.

M.   The difference of the bitter Purging Salt from Nitrous and Calcarious Salts.

B.   Of the Difference of the Bitter Purging Salt, from Nitre, and from the Salt of Lime.

6   G.   De Natura Salis Cathartici Amari, Observationes addendae.

M.   Some additional Observations of the Nature of the Bitter Purging Salt.

B.   Some further Observations of the same Bitter Salt.

Part II

G.   De Salis Cathartici Amari Usu.

M.   Of the Use of the Bitter Purging Salt.

B.   [The same].

1   G.   De Salis Carthartici Amari Usu, generalius.

M.   Of the more General Use of the Bitter Purging Salt.

B.   Of the Use of the Bitter Purging Salt in general.

2   G.   De Modo Salem Catharticam Amaram praescribendi.

M.   Of the method of prescribing the Bitter Purging Salt.

B.   Of the way wherein the Bitter Purging Salt is best prescribed.

3   G.   De Salis Cathartici Amari Usu, specialiter: & Imprimis in Ventriculi Morbis.

M.   Of the more particular Use of the Salt, and first of all of its Use in Diseases of the Ventricle.

B.   Of the Use of the Bitter Purging Salt in particular: and first in Diseases of the Stomach.

4   G.   De Salis Cathartici Amari Usu, in Intestinorum Partiumque adjacentium Morbis.

M.   Of the Use of the Bitter Purging Salt in Diseases of the Intestines and Parts adjacent.

B.   Of the Use of the Bitter Purging Salt in Diseases of the Gut and the parts adjacent.

5 G. De Salis Cathartici Amari usu, in Morbis Cephalicis.
   M. Of the Use of the Purging Salt in Cephalick Disease.
   B. Of the Use of the Bitter Purging Salt in Diseases of the Head.

6 G. De Salis Cathartici Amari Usu, in Morbis quibusdam aliis.
   M. Of the Use of the Bitter Purging Salt in some other Diseases.
   B. [The same].

7 G. De Aquarum Amarescentium, Salisque earum Abusu.
   M. Of the Abuse of the Bitter Waters and their Salts.
   B. Of the Misuse of the Bitter Purging Salt.

(A few of the text captions differ slightly from the headings in the Contents list.)

4   *Sal Catharticus*   Second Edition, 1698

[In a double rule]: Tractatus... Amara [as in 1695], in Aquis Ebeshamensibus, et Hujusmodi Allis Contenti, Natura & Usu. [rule] Editio Secunda. [rule] Authore Nehemia Grew M.D. Utriusque Regiae Societatis Socio. [rule] Londini, Typis Johannis Darbei, 1698.

COLLATION: 12⁰: A⁶, B–E¹²; 6 leaves, 96 pages. Page height 145 mm.

CONTENTS: A1ᵇ Imprimatur to Smith and Walford, as in 1695; A2ᵃ Title-page; A3ᵃ–4ᵇ Dedication, as 1695; A5ᵃ–6ᵇ Elenchus Capitum [without Errata note]; pp.1–96, text as in 1695. Running headings as in 1695, except on p.96: De Salis, &c.

NOTE: The text was reprinted line for line from the first edition in a slightly larger type, which includes some swash capitals. Some of the chapter headings were re-aligned, and there are a few variants of spelling and punctuation. At leaf A3ᵇ the words *olim ante Annos quindecim* are qualified by the footnote, below a rule: *Jam fere novendecim.*

Wing G1959A, not in the Term Catalogue.
COPIES: Royal Society of Medicine, London; Royal College of Physicians of Edinburgh (recorded in first printed Catalogue 1767).

5   *Bitter Purging Salt*   Authorised translation, pirated 1700

[In a double rule]: A Treatise of the Nature and Use of the *Bitter Purging Salt*. Easily known from all Counterfeits by its Bitter Taste.

[rule] Written Originally in Latin, by Nehemiah Grew, Doctor in Physick, Fellow of the College of Physicians and of the Royal Society. [rule] And now done into English. [rule] *London*, Printed in the Year 1700.

COLLATION: 8⁰: A–E⁸; 7 leaves, 63 pages, 1 blank leaf. Page height 140 mm.

CONTENTS: A1 blank; A2ᵃ Half-title: A Treatise of the Nature and Use of the Bitter Purging Salt; A3ᵃ Title-page; A4ᵃ Latin Imprimatur of the Royal College of Physicians, dated Maii 3.1695; A5ᵃ–6ᵃ Dedication 'To the Two Famous and Celebrated Societies of Royal Foundation...', with footnote repeated from Bridges's 1697 edition on A5ᵇ: Now near eighteen Years; A6ᵇ–7ᵇ The Contents; pp.1–29 text: [double rule] Part the First; pp.31–63: [double rule] Part the Second.
    Chapter headings and text as in Bridges's 1697 translation.

Wing G1961.
COPY: Medical Society of London (now at Wellcome Institute) Tract 335 (6).

NOTE: A new printing of the authorised translation of Grew's text, omitting the translator's name and without the justificatory preface 'To the Reader', the 'Table of Falsifications' and the 'Postcript Narrative' condemning Moult's plagiarism, with the same anonymous imprint as his pirated edition of 1697. This appears to be a second piracy by Moult.

Josiah Peter: *Truth in opposition to Falshood*   1701

Truth In Opposition to Ignorant and Malicious Falshood: or a Discourse Written to vindicate the Honour, and to assert the Right of Dr Nehemiah Grew, Fellow of the Royal Society, and R. College of Physicians, London; with respect to his Invention for making the Salt of the Purging Waters, called in his Latin Edition thereof Sal Catharticum Amarum. AND to detect the Injuries done to the Publick, as well as to Himself, by obstructing the Health, and endangering the Lives of the King's Subjects; as also by lessening the Foreign Trade, and otherwise, with the many Counterfeit Salts made and sold by interloping Chymists. Grounded chiefly upon the Testimonies Of many Eminent Members of the Royal Society, and the Royal College of Physicians London, the Royal Colleges of Physicians in Edinburgh and Dublin, the Royal Academy in Paris, and the Imperial Academy in Breslaw, with other Eminent and Learned Persons in England, Scotland, Ireland, France, Italy, Prussia, Poland, and Germany. Collected and composed Out of Books, Letters, and other Papers. [rule] By Josiah Peter Gent. [rule] Magna est Veritas & praevalebit. [rule] London, Printed by J.D. for the Author, 1701.

COLLATION: 4⁰: A–H⁴, I²; viii, 60 pages. Page height 20 cm.

CONTENTS: [i] Title-page; iii–v Dedication to 'Thomas [Tenison] Lord Arch-Bp of Canterbury' . . . [signed] Josiah Peter; vi–viii The Principal Heads of the following Discourse . . . [rule] A List of the Eminent Persons whose Testimonies are made use of . . . [At foot of p.viii below a rule:] Advertisement – [Grew's 'offer to give Satisfaction' by experiment in defence of his method of making Salt from the Purging Water]; pp. 1–60 text, beginning below a double rule and caption heading. At p.30 Testimony of Apothecaries, p.34 Testimony of Physicians, p.54 The Conclusion, p.60 Finis.

COPIES: British Library – 2 copies: 1171.h.18(5) and 1416.h.16; not in NUC.

NOTE: The printer was John Darby.

[Double rule border] Cosmologia Sacra: or a Discourse of the Universe As it is the Creature and Kingdom of God. Chiefly Written, To Demonstrate the Truth and Excellency of the Bible; which contains the Laws of his Kingdom in this Lower World. [rule] In Five Books. By Dr Nehemiah Grew, Fellow of the College of Physicians, and of the Royal Society. [rule] London: Printed for W. Rogers, S. Smith, and B. Walford: At the Sun against St. Dunstan's Church in *Fleetstreet*; and at the Prince's Arms in St. Paul's Church-Yard, MDCCI.

COLLATION: f°: π^2, b^2, A^2, $a–e^2$, $B–Z^2$, $Aa–Zz^2$, $Aaa–Zzz^2$, $Aaaa–Zzzz^2$, $Aaaaa^2$, $Bbbbb^1$; 8 leaves, xviii, 372 pages. Page height 31.5 cm.

CONTENTS: Portrait; 1^a Title-page; $2^a–b1^b$ Dedication to the King; $b2^{a–b}$ Epistle Dedicatory to the Archbishops; $A1^a–2^b$ Preface; $a1^{a–b}$ The Heads of the following Discourse; pp.i–xviii The Contents of the Chapters [and] Errata; pp.1–372 Text (p.132 blank).

On the title-page the words 'Comologia Sacra'. 'God', 'By Dr Nehemiah Grew', 'London' are printed in red. The same woodcut surrounds the initial type-capital at the beginning of each 'Book', but printed upside-down at Book 4; there is a different block round the initial of the Dedication. The gathering Ggg, pp.205–208, is wanting, but the text is continuous.

PORTRAIT: *Effigies Authoris/R White ad Vivum delin et sculpsit.* (Wellcome 1220.1): half-length to right, wearing a full wig; in an oval, with floral ornament, on a pedestal, with shield of arms (argent, a fess chevronné sable between three leopard's masks gules); from the oil-painting in the possession of the Worshipful Company of Barbers of London.

The Heads of the Discourse

Book I: Sheweth that God made the Corporeal World, And what it is.

Chapter 1: Of God. pp.1–5.
　　　　　2: Of the Corporeal World. pp.6–11.
　　　　　3: Of the Principles. pp.11–17.
　　　　　4: Of Compounded Bodies. pp.17–23.
　　　　　5: Of their Use. pp.23–30.

Book II: Sheweth That there is a Vital World which God hath made. And What it is.

1: Of Life. pp.31–36.
2: Of Sense. pp.36–40.
3: Of Mind. And first of Phancy, or Phantastic Mind. pp.41–48.
4: Of Intellectual Mind. pp.48–52.
5: Of the Three chief Endowments of Intellectual Mind. And first, Of Science [Mathematics]. pp.52–56.
6: Of Wisdom. pp.56–62.
7: Of Virtue. pp.62–78.
8: Of Celestial Mind. pp.78–84.

Book III: Sheweth That God governs the Universe he hath made. And in what Manner.

1: Of the Nature of God's Government, or of Providence. pp.85–92.
2: Of the Ends of Providence. And first, in this Life. pp.92–105.
3: Of Providence over Publick States. pp.105–114.
4: Of the Celestial Life. pp.115–120.
5: Of the Rule of Providence. And first, Of the Law of Nature. pp.121–126.
6: Of Positive Law. pp.127–131.

Book IV: Sheweth, That the BIBLE, and first, That the *Hebrew* Code, or Old Testament, is God's Positive Law.

1: Of the Integrity of the *Hebrew* Code. pp.133–43.
2: Of the Truth and Excellency of the *Hebrew* Code. And first, as they appear from Foreign Proof. pp.144–161.
3: Of the Truth and Excellency hereof, as they appear in it Self. And first, if we consider the Writers. pp.162–183.
4: Of the Contents hereof. And first, Of the History. pp.184–193.
5: Of the Miracles. pp.194–204. [No pages 205–208.]
6: Of Prophecies. pp.209–227.
7: Of the Laws. And first, Of those given to *Adam* and *Noah*. pp.227–235.
8: Of the *Mosaic* Law. pp.235–278.

Book V: Sheweth, That the New Testament, is also God's Positive
Law.

  1: Of the Integrity of the New Testament. pp.279–292.
  2: Of the Truth and Excellency hereof. And first, as they
     appear from the Writers. pp.292–310.
  3: Of the Contents. And first, Of the Miracles. pp.311–320.
  4: Of the Doctrine. And first, Of the Revelations we are to
     Believe. pp.320–331.
  5: Of the Laws. pp.332–351.
  6: Of our Saviour's Prophecies. pp.351–372./FINIS.

'The Contents of the Chapters' (pp.i–xviii) gives detail of the numbered paragraphs in groups.

   Term Catalogue, June 1701 (TC III, 255).
COPIES: British Library – 2 copies: 10.c.3, and C.44.g.1 with S.T. Coleridge's autograph marginalia (see below); Royal College of Physicians;
Royal College of Surgeons of England: signature on fly-leaf *Joseph
France, Pr.o–8–o.*; Royal Society, inscribed at foot of title-page *Given by
the Author July 30, 1701*; Wellcome Institute, London. Cambridge University Library N.14.9 Bishop John Moore's copy, *Munificentia Regia
1715* bookplate on title verso; Magdalene College, Samuel Pepys's copy
(*Catalogue* 1978, no. 2304). Bodleian Library, Oxford –Fol.θ 601, early
18th-century inscription on flyleaf *Guift of Cos. Tho. Morse 4d Nath
Symonds* and on title-page *Wm Borlase 1738* with his armorial bookplate
lettered *Wm Borlase Rector of Ludgvan F.R.S.*. Reading University Library, Cole 1969, no.787. Reigate, St Mary's Parochial Library, CR81,
inscribed *This book belongs to the publick Library of Rygate in Surrey Given
by the Learned Author Dr Grew Decr 3d 1703*; Dublin, Trinity College – 3
copies. Edinburgh: National Library of Scotland; Royal College of
Physicians – 2 copies; University Library. Paris – Bibliothèque
Nationale. United States – *NUC–NG '59* records 21 copies including
National Library of Medicine; Harvard University; Institute of the
History of Medicine, Baltimore; Library Company of Philadelphia;
Yale University, Osborn Collection – John Locke's copy (J.R. Harrison
The Library of John Locke 1965).

NOTE: Coleridge's marginalia, in the British Library's copy listed above,
include a long comment signed '*S.T.C.*' about notes which seem to be
by his friend and literary executor the surgeon Joseph Henry Green

P.R.C.S. whose ownership signature is in the book with the bookplate of an earlier surgeon John Pearson, like Green an F.R.S. Most of Coleridge's notes, which occur on 26 pages, are in Books I and IV, with a long prayer on the flyleaf facing the last printed page. The publication, begun in 1980, of Coleridge's *Marginalia* in the Princeton *Collected Works* ('Volume 12') in alphabetic order of annotated authors had been delayed, when I examined this copy in 1985, by the death of the editor George Whalley of Toronto in 1984; but the second volume which includes these *Cosmologia Sacra* notes has now come out (Routledge & Kegan Paul 1987).

John Evelyn's copy was sold in 1977 at Christie's auction room.

The detail of the Reigate copy was generously given me by David Williams, Librarian of the Council for Places of Worship; St Mary's Library was founded in 1701 by the Rector, Andrew Cranston who probably wrote the inscription. (See also Library Association, Library History Group *Newsletter* 1 (1987), 8–10).

VII   Contributions to Philosophical Transactions

1   *The Nature of Snow*   1673

Manuscripts:

Royal Society Archives: (1) Letter to Oldenburg, 12 March 1671/2; (2) R.B.C. 3, pages 267–71.

Printed texts:
(1) Some Observations touching the Nature of Snow. *Philosophical Transactions* 8, no 92 (25 March 1673), 193–96.
(2) Oldenburg's *Correspondence*, ed. A.R. & M.B. Hall, 8, letter 1921.

2   *The Pores in the Skin of Hands and Feet*   1684

Manuscript: Royal Society Archives: R.B.C. 6, pages 87–88.

Printed text: The description and use of the Pores in the skin of the hands and feet, by the learned and ingenious Nehemiah Grew M.D. *Philosophical Transactions*, 14, no.159 (May 20, 1684), 566 [bis]–67[bis], with Figures 1–2 on an unnumbered Plate signed *M Burg, sculp.* [Michael Burghers], facing page 559 [bis].
Phil. Trans. no. 158 ended with page 587 (its verso being blank), but no.159 began with page 559 (in error for 589) and so forward to p.587 [bis].
Figure 1: the palm of the left hand, the fingers pointed upwards, showing in natural size the 'spherical triangles and elliptics' of the skin ridges; figure 2: a diagram of one triangle of ridges showing the pores approximately 3 mm in diameter.

3   *De Morboso Liene*   1691

Manuscript: Royal Society. CP 12 (Anatomy & Surgery) (1) 35: Observationes [etc., as printed] in Grew's hand; note of authorship and date of communication in another hand. 1 leaf (folio height).

Printed text: Observationes aliquot rariores de morboso Liene, a spectatissimo Domino D. Nehemia Grew M.D. ac R.S. Socio cum eadem Societate communicatae Maij 13° 91. *Philosophical Transactions* 17, no.194 (July, Aug., Sept. 1691) 543–44.

4   *The Food of the Humming Bird*   1693

A Query concerning the Food of the Humming Bird; occasioned by
the Description of it in the *Transactions*. Numb 200. *Philosophical Transactions* 17, no.202 (July & August 1693), 815.

5   *The Number of Acres in England*   1711

Manuscript: British Library. Lansdowne MS 691, ff.2a–116b: The
meanes of a most Ample Encrease of the Wealth and Strength of
England In a few Years. Humbly Represented to Her Majestie in the 5th
Year of Her Reign [8 March 1706–7 March 1707].

4⁰: rebound and all leaves mounted; page height 17.5 cm. Text in a
scribe's formal hand.

CONTENTS: 1a (flyleaf) notes concerning Nehemiah Grew in a late
eighteenth–century hand. 2a title; 3a–5b address 'To the Queen's Most
Excellent Majestie', with Grew's signature, not autograph; 6a–8b The
Contents [see below]; 9a–11b text; f.10: folding MS map of England
and Wales, South at the top, lettered by the same hand as the text, only
So. Foreland, New Haven, and *Berwick* named, with letters A–H at the
angles and intersections of the triangulations.

Chapter headings:

The First part: Of improvements upon the lands.
    Chapter 1: Of the dimensions of England and Wales.
             2: Of English Mineralls.
             3: Of the Roads & Rivers.
             4. Of Inclosures.
             5: Of the Culture of the Lands.
             6: Of the Growth upon the Lands.
             7. Of the stock of Cattle and other Usefull Animalls.

The Second Part: Of Manufactures.
    Chapter 1: Of the Manufactures we now have.
             2: Of the Manufactures wch we want.
             3: Of the means of improving our Manufactures.

The Third part: Of improvements by Sea.
    Chapter 1: Of the present condition of our Merchandise.
             2: Of the Advantages we have and may have for im-
               proveing (!) our Merchandise.

3: Of improving our Merchandise by making it more regular.

4: Of improving our Merchandise by making it more Easie.

The Fourth Part: Of the People.

Chapter 1: Of the Means of applying them to the aforesaid Improvements.

2: Of the Means of Multiplying them for the speedier Progress in the aforesaid improvements.

The Conclusion.

Printed Summary:
The number of Acres in England and the Use which may be made of it. *Philosophical Transactions*, 27, no.330 (1711), 266–69, with Tab. 1 a folded map, North at the right, 265 mm. square.

VIII ENQUIRIES RELATING TO NEW ENGLAND AND THE INDIANS 1690

Manuscripts in The British Library

SLOANE MS 4062, ff.235–36: Samuel Lee. A.L.S. 25 June 1690. Folio sheet, written on all four sides. Pp.1–3 text: 118 numbered paragraphs in two columns, p.4: address; the upper corners are defective.
Begins: I have sent you such replies as I can collect in this . . . World to yor Questions: praying yr candid acceptance.
Ends: June 25.1690 at Mount Hope. I rec'd most of the intelligence From one Mr Arnold a practitioner in physick of good request in Rhode Island . . . Sa. Lee.
Address: For the very Learned Doctor Nehemiah Grew M.D. at his Lodging in Fleet Street London. Capt. Saywell I pray inclose it in yrs when you write to him with my service.
Endorsed in Sewall's hand: Recd these Observations Jan^r 24.1680/1.

SLOANE MS 4067, ff.140–41: Samuel Sewall. A.L.S. [2 February 1691]. Large fo. sheet, pages 1–2 text, 3 blank, 4 address. The date was assigned by Kittredge from Sewall's Diary.
Begins: recd ye Revd Mr Lee's Observations . . .
Written across the text: Two more ships are arrived yt sailed from Plim^o Xr.19. bringing supplies of Arms and Ammunition.
Ends: My humble service to Doctor Grew of Coventry, if living. I am Sir, your humble Servt Samuel Sewall.
Address: For Dr Nehemia Grew in Racket Court near Shoe-Lane in Fleet-Street London.
Modern pencilled note in margin: See Sewall's diary for 2 Febr. 1690/1.

These letters were described by George L. Kittredge at a meeting of the Colonial Society of Massachusetts at Boston on 29 February 1912, and printed in his paper in the Society's *Publications*, 14 (1913), 142–86, Lee's letter at pages 145–53 and Sewall's at pages 153–55. Obadiah Grew, to whom Sewall sent his greeting 'if living', had died in 1689. I heard of these letters from my friend Dr Genevieve Miller of Cleveland, Ohio, who quoted Kittredge's paper in her contribution to the Festschrift for Charles Singer *Science, Medicine and History*, 1(1953), p.32, n.14.

IX   ACCOUNT OF HENRY SAMPSON

The Worthy Dr Grew's Account of this His Excellent Brother-in-Law
    In John Howe *A Discourse relating to the Expectation of Future Bless-edness*. London, S. Bridge for Thomas Parkhurst, 1705. 8⁰: 91 pages.

    At p.77: An Appendix Containing some Memorial of Dr Henry Sampson, A late Noted Physician in the City of London. [Includes at pp.79–80 Sampson's farewell letter on his retirement].
    At pp.88–91: Grew's 'Account'.

COPY: British Library 1509/4585.

NOTE: The Account was reprinted in later editions of Howe's *Works*.

X   CORRESPONDENCE

Most of the extant Letters from Grew's Correspondence are in the Royal Society's Letter Books, the British Library's Sloane Manuscripts, and the Bodleian Library's Lister Manuscripts. A few others, with some from the main groups, have been described in the first part of this book at appropriate points.

Royal Society: Early Letters, Guard-Book 8 includes official letters, many in Latin, sent or received by Grew as Secretary of the Society between autumn 1677 and February 1680, though he had nominally resigned in November 1679. These Society correspondents included Johann Bohn (Liepzig), Ismael Bouillau (Paris), Martin Fogel and Georg Held (Hamburg), Johann Hevelius (Danzig), Christian Huygens (The Hague), G.F. Leibnitz (Hanover, etc.), Marcello Malpighi (Bologna), Christian Mentzel (Berlin), Francesco Nazzari (Rome), R.F. Sluse (Liège), and Jan Swammerdam (Amsterdam). Letters to Grew from Sir Robert Sibbald (Edinburgh 1683), Solomon Reisel (Stuttgart 1684), and Robert Thompson (Port Royal 1696) are in later volumes.

Letters from and about Grew are printed in A.R. & M.B. Hall's edition of Henry Oldenburg's *Correspondence*; the strayed letter to Oldenburg discovered by Dr Michael Hunter has been described above (p.2) in my account of Grew's appointment as Curator at the Royal Society. Grew's correspondence with Malpighi has been published in full by H.B. Adelmann; Leeuwenhoek's letters, partly published in the *Philosophical Transactions* at the time they came in, were discussed in detail by Clifford Dobell (See VIIA, page 57 above). A letter from Grew introducing Sir Thomas Molyneux of Dublin to Robert Plot at Oxford was reprinted from the *Dublin University Magazine* 18 (1841), 324 by C.H. Josten in his *Elias Ashmole*. (Oxford 1966).

British Library: Letters to and from Grew are included in several Sloane MSS: from Grew to Sloane himself in vols. 2145, 4036, 4039, 4042, 4059; there are letters from Sir Thomas Browne (MS 1912), Louis LeVasseur (MS 1926), Petty and Wren and J.B. De la Roque (MS 1942), Ralph Thoresby (MS 4025), with others concerning his researches in MSS 4037, 4076 and 4811. Letters of personal interest or connected with his medical practice are in MSS 2145 and 4066. For his letters from New England friends in MSS 4062 and 4067 see Chapter VIII.

Bodleian Library, Oxford: Lister Manuscripts 3 (f.111) and 34–35 include many letters from Grew to Dr Martin Lister between 1671 and 1682, with letters to Lister concerning Grew from Henry Oldenburg (1671–76), Sir John Brooke (1672), and Francis Aston (1682).

NOTE: Josiah Peter *Truth against Falshood* 1701 prints numerous letters answering Grew's circular enquiry of 1696 about the use of his Epsom Salt by physicians and its efficacy compared to that of 'counterfeit' Salts, for which see V, 6.

[In a black border] Enoch's Translation [rule] A Funeral Sermon upon the sudden death of Dr Nehemiah Grew Fellow of the College of Physicians. Who died March 25th, 1712. Preach'd at Old-Jewry. [rule] By John Shower. [rule] London: Printed by J.R. for John Clark, at the Bible and Crown in the Old Change; and may also be had of Eb. Seadgel, at the Tea-Canister in Bartholomew-Close. 1712.

COLLATION: 8^0: π^2, B–C^4, D^3; 2 leaves, 22 pages. Page height 18 cm.

CONTENTS: 1^a Title-page; 2 Dedication To Sir Richard Blackmore, Kt. May 10, 1712; pages 1–22 The Sermon, on the text: Gen.v.24 'Enoch walked with God, and he was not, for God took him'.

COPIES: British Library; Royal College of Physicians; Royal College of Surgeons of England; Cambridge University Library; Bodleian Library, Oxford. United States (NUC–NS0520853): Harvard; W.A. Clark Library, Los Angeles; Yale.

Appendix I

Donors to the Royal Society's Repository, and Subscribers to
Musaeum 1681 or to *The Anatomy of Plants* 1682

Three lists of names were printed in connection with Grew's two large books of 1681 and 1682. The Prospectus for each included the names of the earliest subscribers, and on the last page of the Catalogue of the Royal Society's Repository there is 'A List of those who have contributed to this *Musaeum*; excepting some Names which are lost'. These lists are of historical importance in naming the Fellows and those from outside the Society who supported its work by gifts or subscriptions.

A dozen donors named in the text of *Musaeum* were left out of the List, while five of those listed are not recorded in the text. Further subscribers to the books are known from other records: eighteen from Yorkshire and Nottinghamshire for *Musaeum* were named by Grew in his letter of 18 May 1681 to Dr Martin Lister at York, and three from Norfolk to *The Anatomy of Plants* by Sir Thomas Browne in his letter of 29 May 1682 to his son Dr Edward Browne in London. The Yorkshire copies were sent in 1681 by Moses Pitt, the London bookseller, to Richard Lambert, bookseller at York; Grew told Lister in his letter of 3 September 1682 that Lambert had delivered 16 of the 18 copies sent to him, paid Grew ten pounds, and ordered more copies. Lambert was in business till 1686, perhaps till 1690 (H.R. Plomer *A Dictionary of Printers and Booksellers 1668–1725*, 1922).

Daniel Colwall's name is not in these lists, but his constant support was gratefully acknowledged in the Dedication to him of *Musaeum* and by the frontispiece portrait there on which he is called *Musaei Regalis Societatis Fundator*; he had made two gifts of £50 each in 1665 for the Repository, and now added a 'Voluntary Undertaking for the Engraving of the Plates' in Grew's Catalogue of it.

William, [5th] Lord Paget died in 1678; the subscriber to *musaeum* listed in the Prospectus (1680) was perhaps Thomas, 6th Lord Paget.

The names in these lists have been gathered here into a single alphabet, and their spelling regularised as far as possible. The letter or letters following each name show if the man was a Fellow of the Royal Society and in which list his name appeared, by these initials:

D   Donor to the Museum.
F   Fellow of the Royal Society.
L   in the list of Donors but not in the text of *Musaeum*.
M   Subscriber to *Musaeum*.
N   Norfolk subscriber.
P   Subscriber to *The Anatomy of Plants*.
T   in the text of *Musaeum* as a donor, but not in the List.
Y   Yorkshire subscriber.

Prince Maurice *D*
Prince Rupert *F, D*
Sir Edward Abney *M*
Aerskine *see* Erskine
African Company *see* Royal
William Aldworth *P*
Dr Thomas Allen *F, D, M, P*
Arthur Annesley, Earl of
    Anglesey *F, M, P*
Elias Ashmole *F, P*
Francis Aston *F, M, P*
John Aubrey *F, D*
Sir William Ayscough *Y, M*
Dr W. Ayscough *Y, M*
John Baker *M*
Thomas Bard *M*
Samuel Barnardiston *M, P*
Thomas Barrington *F, M*
Erasmus Bartholin *D*
Richard Baxter *M*
John Beilby *Y, M*
William Bell *M*
John Bemde *F, D*
Robert Bennett *M*
George, 1st Earl of
    Berkeley *F, M, P*
Dr Francis Bernard *P*
Dr John Betts *P*
Robert Blaney *M*
Paolo Boccone *D*
Henry Bokenham *N, P*
John Boothby *M*
Olaus Borrichius *D*
Joseph Bowles *D*
Ralph Box *P*
Hon. Robert Boyle *F, D, M,*
    P
John Egerton, Viscount
    Brackley *M, P*
John Egerton, 2nd Earl of

Bridgewater *M, P*
Dr William Briggs *M, P*
Dr Humphrey Brook *P*
Sir John Brooke *F, Y, M*
William, Viscount
    Brouncker *F, D, P*
Dr Edward Browne *F, D,*
    M, P
Sir Thomas Browne *D, N, P,*
William Brownest *D (T)*
Dr Anthony Burgess *M, P*
Dr William Burnet *M*
Hezekiah Burton *M*
Archbishops of Canterbury,
    see Sancroft, Tenison and
    Tillotson
Dr Nicolas Carter *P*
Dr Walter Charleton *F, D, P*
Bishop of Chester, *see* Wilkins
Walter Chetwynd *F, D*
Henry Hyde, 2nd Earl of
    Clarendon *F, P*
Samuel Clark *D (T)*
Peter Clayton *M*
Sir Robert Clayton *M*
Dr Andrew Clench *F, D, M,*
    P
Dr Josias Clerk *P*
Samuel Colepress *D*
Dr Samuel Collins *P*
Daniel Colwall *F, M*
Henry Compton, Bishop of
    London *M*
Anthony Ashley Cooper, *see*
    Shaftesbury
Dr Edward Cotton *F, D*
Daniel Coxe *F, M, P*
Thomas Coxe *F, D, M, P*
Dr John Creed *F, P*
Ellis Crispe *D*

Henry Crispe *M*
Sir Thomas Crispe *F, D (L)*
Dr William Croone *F, D, M, P*
'Henry, Earl of Daventry', *see* Heneage, Lord Finch of Daventry
Dr William Dawkins *P*
Dr Thomas Dawson *P*
Sir Theodore de Vaux, *see* Vaux
Thomas Day *F, M*
Dr John Downes *F, M*
John Dubois *M*
East India Company *D (L)*
Dr Francis Edes *M*
Egerton, *see* Brackley, and Bridgewater
Ely, *see* John Moore, Bishop
Dr George Ent *F, D, M, P*
William Erskine *F, M, P*
John Evelyn *F, D, P*
Dr Nathaniel Fairfax *D (T), M*
Dr John Feake *M*
Heneage, Lord Finch of Daventry *M*
Thomas Firmin *F, M, P*
Thomas Fissenden *D*
Fitzgerald, *see* Kildare
James Fraser *P*
Frederick of Holstein, *see* Holstein
Percy Freke *M*
Thomas Gale *F, M*
Dr Charles Gibson *P*
Jonathan Goddard *F, D (T)*
Dr Charles Goodall *M*
Lovett Goring *M*
Peter Gotte *M*
Dr Nehemiah Grew *F, D*

Theodore Haak *F, D, M*
John Hampden *M*
Hastings, *see* Huntingdon
Dr Hawy *N, P*
Edward Hearne *P*
Thomas Henshaw *F, D, M, P*
John Herbert *F, M, P*
Sir Thomas Herbert *Y, M*
Sir John Hewley *Y, M*
Abraham Hill *F, D, M, P*
....Hocknel *D*
Dr Luke Hodgson *D*
William Holder *F, M, P*
Francis, 2nd Viscount Holles *M*
Friedrich, Duke of Holstein *D (T)*
Robert Hooke *F, D, M, P*
Anthony Horneck *F, D, M, P*
Sir John Hoskins *F, D*
John Houghton *F, D, M, P*
Charles Howard *F, D, M, P*
Henry Howard, *see* Norfolk *Duke*
Lord Thomas Howard *F, P*
Dr Howman *N, P*
Dr Edward Hulse *M, P*
Thomas Hastings, 7th Earl of Huntingdon *M*
Henry Hyde, *see* Clarendon
Thomas Jacomb *M*
Johann–Georg, *see* Saxony
Dr Samuel Jordan *M*
John Kaye *Y, M*
John Fitzgerald, 18th Earl of Kildare *M, P*
Kilmore, *see* Francis Marsh, *Bishop*

Dr Edmond King F, D, M, P
Richard Lambert Y, M
Joseph Lane F, M, P
Lucas Langerman D
Thomas Langham M
Benjamin Lannoy D
Sir John Lawrence F, P
Sir John Legard Y, M
William Lightfoot M
J. Linger D
John Lisle Y, M
Robert Sidney, Viscount Lisle M, P
Christopher Lister Y, M
Dr Martin Lister F, D, Y, M
Geronimo Lobo D
London, Bishop of, see Compton
Thomas Long M
Christopher Love-Morley M, P
Dr Richard Lower F, D
Sir John Lowther F, M, P
John Malling D
Marcello Malpighi F, D
John Marlow M
Francis Marsh, Bishop of Kilmore M
Robert Marshall Y, M
Dr John Master P
Maurice, Prince, see at head of list
Dr Christopher Merrett F, D, M
Sir John Micklethwaite M, P
Sir Thomas Millington D, M, P
Daniel Milles F, M, P
Dr Walter Mills F, P

Samuel Moody D (T)
John Moore, Bishop of Ely M, P
Sir Jonas Moore F, D
Jonas Moore F, M
Sir Robert Moray F, D
Dr Richard Moreton M, P
Samuel Morgan D
Morley, see Love-Morley
John Morris M
Roger Morris M
William Moses M, P
Christopher Montague P
Willam Napper F, M, P
Dr Walter Needham F, D, M, P
Sir Paul Neile F, M, P
Robert Nelson M
Sir Isaac Newton F, D
John Nicoll M, P
Michael Noble M
Henry Howard, 6th Duke of Norfolk F, D
Sir Thomas Norton M
Dr Thomas Novel F, M
Henry Oldenburg F, D
William Osbaldeston Y, M
Philip Packer F, D, M, P
William, Lord Paget M
Dudley Palmer F, D
John Parsons M
William Payne F, P
Samuel Pepys F, P
William Perkins M
Sir William Petty F, D, M
Sir Thomas Player F, M
Robert Plot F, D (L)
Dr Walter Pope F, D
Dr Thomas Povey F, D, M
Henry Powle F, D (T)

John Rawlet P
William Rawlinson M
Sir Robert Reading F, M, P
Sir Metcalfe Robinson M
Dr Tancred Robinson F, M
Thomas Rokeby Y, M
John Rosse Y, M
John Rosseter M, P
Royal African Company D
  (L)
Luke Rugeley P
James Ruffine P
Rupert, *Prince, see* at head of
  list
Sir Philip Rycaut F, P
Salisbury, *see* Seth Ward,
  Bishop
Dr Henry Sampson M, P
William Sancroft, Archbishop
  of Canterbury M, P
Samuel Sanders M
Clement Sankey M, P
Johann-Georg, Duke of
  Saxony D (T)
Sir Charles Scarburgh F, M
Joshua Scottow D
Dutton Seaman M
Anthony Ashley Cooper, 1st
  Earl of Shaftesbury F, M
John Sharp M
James Shipton P
John Short D
Robert Sidney, *see* Lisle
Dr William Simpson M
Sir Philip Skippon F, D
Dr Frederick Slare F, D, P
Dr George Smith F, D
John Somner D
Sir Robert Southwell F, D,
  M

Edward Stillingfleet P
Dr William Stokeham M, P
Dr Chirstopher Stone Y, M
Samuel Stubbs M
Jan Swammerdam D
Dr Thomas Sydenham M
Captain William Tayler D
Thomas Tenison [Archbishop
  of Canterbury 1694] P
Dr Davis Thomas M
Jeremiah Thompson M
John Tillotson [Archbishop of
  Canterbury 1691] F, M, P
Dr John Torkington M
Dr Richard Torless P
George Trumball D
Sir William Trumbull F, M
Sir William Turner M
Dr Edward Tyson F, D, M, P
Charles Umfreville M
Peter Vanderput M
Sir Theodore de Vaux F, M,
  P
Bartholomew Vermuyden M
Cornelius Vermuyden F, M
Sir Philip Vernatti D (T)
William Vertrey M
William Wagstaffe M, P
Richard Waller F, P
Seth Ward, Bishop of
  Salisbury F, D, M, P
Dr William Warner P
Dr Henry Watkinson Y, M
Samuel Western M
William Wharton M
Joseph Whatmough M
George Wheeler D
Benjamin Whichcote M
Sir Paul Whichcote F, D (T),
  M

Dr Daniel Whistler *F, D, M, P*
Henry Whistler *D*
John White *M*
Michael Wicks *D (T)*
John Wilkins, Bishop of Chester *F, D*
Sir Joseph Williamson *F, D, M, P*
William, 6th Lord Willoughby of Parham *D (L)*
Francis Willughby *F, D*

Sir John Winn *M*
John Winthrop *F, D*
Anthony Wither *M, P*
Dr Thomas Witherley *P*
Dr Robert Wittie *D, P*
Robert Wood *F, P*
Sir Christopher Wren *F, D, M, P*
Dr John Wright *M*
Sir Cyril Wyche *F, M, P*
Dr John Yarborough *Y, M*

Appendix II

Philosophical Transactions – six issues edited by Nehemiah Grew, completing Volume 12 (Term Catalogue, Michaelmas (November) 1679 – TC I, 372)

137. February 10 for January and February 1677/8, pages 923–44.
— John Graves. The manner of hatching chickens at Cairo.
— Sir Robert Moray, A relation of some barnacles; A description of the island Hirta.
— Jonathan Goddard. Some observations of a cameleon.
— Henry Powle. An account of the iron works in the Forest of Dean.
— Sir Philiberto Vernatti. The making of ceruse.
— *Reviews of* Ralph Cudworth *The true intellectual System of the Universe,* part 1, and J.B. Tavernier. *The six Voyages thro Turky into Persia and the East,* 1678.

138. March 25, 1678, pages 945–68.
— Charles Howard. The planting and ordering of saffron.
— Christopher Merret. The tin mines in Cornwall.
— Jonathan Goddard. Experiments of the refining of gold with antimony.
— Dr S. Morris of Petworth. A monstrous birth.
— *Reviews of* Moses Charras *The Royal Pharmacopoeia,* Thomas Hobbes *Decameron Physiologicum,* and Joseph Moxon *Mechanic Exercises.*

NOTE: Grew's autograph draft of the first page (945) of the March issue is in Sloane MS 1942, at f.35.

139. For April, May and June 1678, pages 969–88.
— Ismael Bouillau. Occultation of Saturn by the Moon.
— Robert Boyle. Red snow.
— J.G. Duverney. The structure of the Nose.

—M. Guattoni and D. da Piacenza. Some animals and a strange plant... of the Congo.

—Edmund Pitt. Sorbus pyriformus observed to grow wild in England.

—P. Bayle. A child 26 years in the Mother's belly (from *Journal des Scavans*).

—*Reviews of* John Wallis *De cometarum distantiis*, Martin Lister *Historia animalium Angliae*, and Robert Hooke *Lectures and Collections*.

—Advertisement of Joseph Moxon *Continuation of Mechanic Exercises*, with Roger Palmer, Earl of Castlemaine *A new kind of Globe*.

140. For July and August 1678, pages 999–1014.

—Henry Sampson. Anatomical Observations of a woman who died hydropsical, translated 'out of his Latin' by Grew.

—Antoni van Leeuwenhoek. Microscopical Observations of teeth, bones, ivory and hair.

—G.A. Borelli. The price of his telescope.

——de Gennes. A new clock; A new invention to make linen cloth (from *Journal des Scavans*).

—Matthew Milford. A worm voided by urine.

—George Ent. A conjecture of tempers and dispositions by modulation of the voice.

—*Reviews* (from *Journal des Scavans*) *of* Lorenzo Legati. *Museo Cospiano* (Ferdinando Cospi's Museum joined to Ulisse Aldrovandi's), *Systema Bibliothecae Parisiensis Societatis Jesu*, C.D. Du Cange *Glossarium mediae et infimae Latinitatis*, and Daniel Duncan *Explication des actions animales*.

141. For September, October, November 1678 [imprint 1679], pages 1015–34.

—G.D. Cassini. Lunar eclipse 29 October 1678.

—J.C. Gallet. Solar eclipse 11 June 1676.

—Michael Butterfield. The making of microscopes.

—John Conyers. Experiment of Sir Samuel Morland's speaking trumpet.

—*Reviews of* Thomas Trapham *The state of health in Jamaica,* and Edmund Halley *Catalogus stellarum australium*, with Observation of the transit of Mercury 28 October 1677.

142. For December 1678, January and February 1678/79, pages 1035–74

—Edward Tyson. Anatomical Observations: abscess in the liver, stones in the gall bag, unusual conformations [etc]

— A. van Leeuwenhoek. De natis e semine genitali animalculis, Nov.
  1677; Grew's Latin answer, 1 January 1678; extracts from Leeuwen-
  hoek's letters of 18 March and 31 May 1678.
— Christopher Merret. The art of refining.
— Daniel Colwall. English alum works; English green copperas.
— Thomas Rastell. The salt works of Droitwich.
— John Winthrop. The culture and use of Maiz [in New England].
— Sir Robert Moray. The manner of making malt in Scotland.
— *Review of* George Ent *Antidiatribe, animadversiones in Malachi Thrustone
  Diatribam de respirationis usu.*

Appendix III

'Papers of Dr N. Grew'

Sloane Manuscript 1941 in the British Library is a guard-book containing drafts and fragments in various hands, mainly unsigned and undated,
but many certainly by Grew himself. There is a draft in Latin (f.84) of
the first two paragraphs of *An Idea of a Phytological History*, published in
English in 1673, in a more formal hand than Grew's usual writing, and
two schedules of classified pharmacy terms, 'Essences' etc. (ff.80–81) also
in Latin, similar to the Latin 'Table which includes all the different kinds of
Tastes' in *The Anatomy of Plants* (1682, p.295) and the English 'Schemes' of
shells in *Musaeum* (1681, pp.150–53).

The Certificate of Grew's first marriage on 20 April 1673 to Mary
Huetson is here (f.17), and there are several papers (ff.26–30) connected
with Grew's work as Secretary of the Royal Society in 1677–79: lists of
foreigners 'of literary eminence' to whom letters had been sent, a list of
physicians in Paris (f.23), which however does not include his translators
Louis LeVasseur and Guy Mesmin, fragments of catalogues of books
(f.35), lists (ff.37–40) in alphabetical order of counties from Lancashire to
Nottingham of local medical men, perhaps possible subscribers for
Musaeum or *The Anatomy of Plants*, with a list of English names (f.19)
roughly alphabetised including many Fellows of the Royal Society and
some names of women; 'Mr Hobbs' is in this list, probably Thomas
Hobbes the philosopher who died in 1679.

There is a series of questionnaires, some in English others in Latin, about
medical education, control of practice, and hospitals in the Netherlands,
Germany, Poland, and various Italian cities, with a general one 'of Medicine in Europe' (ff.59–62). The first questionnaire (f.6), about Leiden and
Amsterdam, is headed 'Queries sent to Dr Stanley Chaplaine to ye Prin-
ces[s] of Orange', there are eighteen questions; it must date from the late
1680s, for William Stanley became Princess Mary's chaplain in '85 and she
was proclaimed Queen of England in February '89. These questionnaires

resemble Grew's enquiries about Massachusetts in 1690.

There is also a handwritten pamphlet about the Conformity Bill, which was causing unease among Dissenters between 1701 and its passage into law in November 1710. 'Occasional conformity' was permitted to Dissenters under the Corporation Act of 1661, allowing them to accept official appointments if they had received Communion by the Anglican rite during the previous year. This pamphlet (ff.7–12), entitled *The True Catholic*, argued that 'solemn and sacred' Communion in a Dissenting congregation was of equal spiritual value to that of the Church of England, while divisions within the Church verged on schism. The tone of *The True Catholic* is latitudinarian, but less 'ecumenical' than *Cosmologia Sacra* which Grew, though a devout Presbyterian, dedicated in 1701 to the two Anglican archbishops. After the Restoration many formerly republican 'projectors' became the accepted 'scientists' of the establishment. Charles Webster, reviewing S. Shapin and S. Schaffer *Leviathan and the Air-pump* in *The Times Literary Supplement* 13 March 1987, suggested that the re-established Church preferred the experimental science of the young Royal Society to the materialistic philosophy of the 'atheist' Thomas Hobbes, who saw Robert Boyle as 'the aristocratic instrument of the clientele of the discredited and chaotic republican regime who were covertly forcing their values on the restoration monarch'.

The pamphlet must date from later than 1701, when the Conformity Bill was introduced in the House of Commons, after the accession of Queen Anne, who was firmly Anglican. The hand of the manuscript is probably Grew's, but the text is nearer to the thought of his friend the Presbyterian minister John Shower. When the Bill was passed into an Act of Parliament in 1710 Shower protested to the Lord Treasurer Robert Harley, Lord Oxford; it is thought that his unyielding answer was composed by his 'propagandist' Jonathan Swift, who had watched the progress of the Bill with the interest of a 'high church' clergyman. (Swift *Journal to Stella* 22 December 1711; Irvin Ehrenpreis *Swift* 2 (1963), 144 etc.; Swift *Correspondence*, ed. Harold Williams 5 (1965), Appendix VIII).

For the rest, there is an account (f.78) of the death of James Coates, the Butler at Pembroke College, Cambridge, several contemporary verses and pasquinades in various hands, and a few large folios in shorthand.

General Index

Names in Appendix I are indexed only where they occur in the main text

Academia Naturae Curiosorum *see* Miscellanea curiosa
Académie (Royale) des Sciences, Paris 14, 59, 142
Account of Henry Sampson 68–9, 151
Act of Toleration 53
  of Uniformity 68
  see also Conventicle Act, Five Mile Act
Acton mineral water 49
Adelmann, Howard B. *Marcello Malpighi* and *Correspondence* xv, 8, 10, 30, 70, 87, 152
Agreements and Conditions (Salt-water sweet'ned) 47, 125–6, 129, 131
air-pumps 32–3
Aldrovandi, Ulisse 35, 38, 164
Allen, Benjamin *Natural History of Mineral Waters* 52
amadavat 42
Amerindians 34, 67–8, 71, 150
analogies 11–2, 20, 27, 42, 55
Anatomia Vegetalium inchoata 9, 77
Anatomical Tables, Evelyn's and Harvey's 39
anatomists, comparative 40
  Renaissance 35
  history of, *see* F.J. Cole
anatomy, comparative xi, 21, 27, 35, 41–2, 55
Anatomy and Vegetation of Roots 14–7, 25, 85–8
Anatomy of Plants xviii, 2, 3, 5, 7–27, 30, 36–7, 59, 97–108, 153–61, 167; Plates 3–6
Anatomy of Vegetables Begun 2, 8–16, 21, 25, 77–85
Anne, Queen 65, 148, 168
Appleby, John 36, 51
Arber, Agnes *From medieval Bestiaries to modern Botany* xv
Archilochus alexandra, – Colubris 63
Aristotle 17

Arnold, Caleb, physician in Rhode Island 67, 150
Aromatari, Giovanni *De Generatione Plantarum* 9
Arundel, Earl of, *see* Thomas Howard
Ashmole, Elias 24, 78, 89, 92, 127, 129–30, 135, 152
Aston, Francis 153
astronomy 54, 57
Atherton, Warwickshire 1
atomic theory 18, 55
Aubrey, John 49
Aylmer, B., printer 134

Bagford, John, Collection of Title-pages, etc. 98, 112
Balliol College, Oxford 1
Banks, Sir Joseph 80, 83–4, 86, 91
Barbadoes 132
Barbers Company 143
bark 16, 20
Barnet mineral water 49
bats 35
Bayle, Pierre 164
Bayly, Mr 65
Beale, John 45, 127–30
beaver 35, 43
Bell, John, *Circulating Library Catallogue* 105
Bennett, J.A. 'Wren's last building?' 38
Bennett, Joseph (Lilly Library) 100
Berkeley, George, 1st Earl of Berkeley 131
Bernard, Claude 41
Bibliotheca Norfolciana 38
Birch, Thomas *History of the Royal Society* xv, 2, 7, 56, 76
birch-bark utensils 34
birds 35, 41–2

black plate in *Trunks* 19, 20, 91
Blackmore, Sir Richard 72, 154
Bobart, Jacob, the younger 117
Bohr, Johann 152
Bokenham, Henry 24
Bolam, Jeanne 'The botanical work of
    Nehemiah Grew' xv
Bologna 10
Borelli, Giovanni Alfonso 26, 164
Borlase, William, signature and bookplate 145
Bostock, John 78
'Botanical Papers' 8, 106–07
Bouillau, Ismael 152, 163
Boulger, G.S. 'Nehemiah Grew' xv
box-hive 34
Boyle, Lady Jane, Countess of Kildare 46
Boyle, Richard, 1st Earl of Cork 46
Boyle, Robert 11, 14, 18, 21–2, 24–5, 28, 30,
    32–3, 44–5, 49, 55, 60, 81–3, 85, 96, 122,
    127, 129–30, 132, 163, 168
branches 12, 18–20, 37
Brazil 35
Brewster, Edward, printer 129
Bridge, S., printer 151
Bridges, Joseph 50–2, 137–8, 141
Bridgman, William 46, 131, 133
Brief of Two Treatises 48, 123, 132–3
British Museum 32, 36, 39, 64, 71
British Museum (Natural History) 37, 39, 40
British Union Catalogue of Periodicals 81
Britten, James *The Sloane Herbarium* 107
Brooke, Sir John 153
Brouncker, William, Viscount 2, 13, 19, 24,
    34, 56, 58, 89, 116
Brown, Robert, botanist 16
Browne, Edward 24, 32, 105, 155
Browne, Sir Thomas 12, 18, 24, 105, 152, 155
Brunner, Johann Conrad *Experimenta nova* 62
buds 26–7
Buffon, Georges Louis 64
bulbs 27
Buonanni, Filippo 26
Burgess, Renate *Portraits of Doctors and
    Scientists* xv, quoted 143
Burghers, Michael – 'M. Burg', engraver 147
Burnet, Gilbert, Bishop of Salisbury 22, 53
burning-glasses 33
Butt, Edward, of Wimborne, ownership

signature 120
Butterfield, Michael 164

Cademan, William, printer 127–8, 130
Calceolari, Francesco 38
calculi 21
Calamy, Edmund *Ejected Ministers* 69
Cambridge xi, 1, 3, 8, 9
Carruthers, William 'The Life and Work of
    Nehemiah Grew' xv
Cary, Anthony, 5th Viscount Falkland 131,
    133
Cassini, G.D. 164
cassowary 41
Castlemaine, Earl of, *see* Roger Palmer
castor, castoreum 35
*Certificates of several Captains (Saltwater
    sweet'ned)* 47, 126, 129, 133
Chamblet, Sam. 132
Chancery Lane pump 44
Charles II 19, 28, 44–6, 51, 124, 129, 132, 143
Charras, Moses 163
Chartered Trading Companies 107
Chatto, John 94
Chauveau, F., engraver 79
Cheapside conduit 44
Chelsea College 59
chemistry 7, 8, 17, 18, 21, 23, 26, 75
Cheney, Mr 59
Cheshunt, Hertfordshire: parish records 3
Chester, Bishop of, *see* John Wilkins
chimpanzee 61
Chinese: ginseng root 36; observation of
    snowflakes 60
Chiswell, Robert, printer 15, 86
chocolate, eaten with maize 34
Christ Church parish, London 72
Christ's Hospital pump 44
Church of England 1, 53, 68, 168
cider press 34
'Circular Letter to Foreign Correspondents'
    (Royal Society) 56
Clark *see* Lewis and Clark
Clark, Sir George 72
Clark, John, printer 152
classification 15, 35–7, 41, 61
Clayworth, Nottinghamshire 68
Clifford, Abraham, minister and physician 5

Clench, Andrew 36
Clift, William, Conservator of John Hunter's Museum 39
Coates, James, butler of Pembroke College, Cambridge 168
Cobham water 49
Cohen, I. Bernard 106
Colbert, Jean-Baptiste 80
Cole, Dennis 72
Cole, Francis Joseph xv, 19, 35, 40–1, 69; *see also* Index of Libraries: under Reading University
Cole, William, physician 42
Coleridge, Samuel Taylor, marginalia 53, 145–6
Collins, Samuel 135
Colonial Society of Massachusetts 150
colours of plants 25
Colwall, Daniel xvii, 28, 30, 112, 116, 155, 165, Plate 7
Comparative Anatomy of Stomachs and Guts xvii, 7, 28, 38, 40–3, 111–3, 115, Plate 2
Comparative Anatomy of Trunks 18–20, 25–6, 61, 90–3
Compton, Henry, Bishop of London 35
condensing engine: Boyle's 33, Papin's 90
Conditions for the Use of the Invention (saltwater sweet'ned) 130–1
Conformity Bill 1701–10 168
Conventicle Act 1664 1, 68
Cony, Robert, bookplate 96
Conyers, John 164
Cork, Earl of, *see* Richard Boyle
Corporation Act 1661 168
Cosimo III, Grand Duke of Tuscany 116
Cosmologia Sacra xvii, 53–7, 143–6, 168, Frontispiece
Cospi, Ferdinando 38, 57, 164
cotyledons 12
Coventry 1, 3, 7, 14, 63, 150
Cowper, William, anatomist 39
Coxe, Daniel 18
Crane Court, London 38–9
Cranston, Andrew 146
Crauford, Robert 47, 132
Creagh, Andrew 130
creeping plants 20
Crispe, Thomas 39

Crist, Timothy J. 75
crocodile 39
Croone, William 32
Cruikshank, William 61
Cudworth, Ralph 163
Culpeper, Nicholas 94
Curtis, William 105

dactyloscopist 61
Dandy, J.E. *The Sloane Herbarium* 107
Danzig, 54, 56
Darby, John, printer 137, 140, 142
Darwin, Charles 16
Dawson, Warren R. *Catalogue of Mummies* 36
Dedu, N. 11, 81–2, 85
de Graaf, Regnier 42
Deleboë (Dubois, Sylvius), Franz 5
demonstrations to Royal Society 2, 7, 11, 49
Descartes, René 60
Deutsche Akademie der Naturforscher 82
development of plants 33
Dezallier, A., printer 27, 80
diabetes 52
digestion 41
Discourse concerning Mixture 17–8, 25–6, 88–90
discourses to Royal Society 7, 10, 12, 14–6, 19, 23, 25, 27, 36, 44, 54, 122
Disputatio de Liquore nervoso 5, 76
distance-meter ('waywiser') 33
Dix, E.R.M. *Dublin-printed Books* 127
Dobell, Clifford *Antoni van Leeuwenhoek* 57–8, 152
Dodart, Denis *Mémoires . . . Histoire des Plantes* 51, 59
Doddington, George 131
Dodson, Elizabeth, Mrs Grew 3
Dodwell, Henry, signature 86
donors to Royal Society's Museum 30, 112, 155–61
Douce, Francis, books and bookplate 92, 117
dredging machine 33, 65
Dryander, Jonas 80, 86
Dublin: Archbishop 53, physicians 51, printing 47, 127
Dublin University Magazine 152
Dubois *see* Deleboë
Du Cange, C.D. *Glossarium* 164
Du Hamel, Jean-Baptiste *Philosophia vetus et*

nova 14
Dulwich water 49
Dun, Sir Patrick 51, 70
Duncan, Daniel 164
Du Verney, J.G. 163

Eales, N.B. *The Cole Library . . . Catalogue* xv
East Indies 34–5
echeneis, see ship–halter fish
economic project 33, 64–6, 148–9
Edwards, W.N. *The History of Palaeontology* 37
eggs: of cassowary 41, double 29, of guillemot 37
electricity 55
elephant tusk, wreathed 39
Elzevir, printers 5, 76
emu 41
'Enquiries relating to New England and the Indians' 67–8, 150
Ent, Sir George 22, 164–5
Epsom: salt, spa, waters xvi, 49–52, 135–42, 153
'Essential and marine Salts' 25
Europe: medical institutions, practice, etc. 167
Evans, Thomas, Fellow of Eton 117
Evelyn, John 9, 24–5, 51, 53, 70, 91–2, 105, 117–8, 146; 'Anatomical Tables' 39, 118; Bibliography 39; *Diary* quoted 10, 39, 49, 116–8
Ewen, A.H. *Flora* by John Ray edited 9
exotic artefacts 34; – plants 15
Experiments in Consort of Luctation 10–11, 17–20, 23, 25, 93–97, 135
eye model 33

Fabri, Honoré *De Plantis* 9
facsimile editions 20–1, 26–7, 95, 105–6
Falkland, Viscount *see* Anthony Cary
feather mantle and perruque, Amerindian 34
feeding habits: mammals 40–1, humming-bird 62–4, 148
Fetter Lane, London 38
Finch, Charles 4th Earl of Winchelsea, bookplate 120
Finger-print whorl'd 61
Fitzgerald, George, 16th Earl of Kildare 46
  Robert 44–8, 73, 122–3, 125–7, 131–3

Five-mile Act 1665 3
'Fixed Salts, Discourse' 36
Flamsteed, John 105
Fleet River 72
  Fleet Street 38, 72
  floor, Wilkins's geometric 33
florets in compositae 12
'Flowers, Discourse' 12, 25, 27
Floyer, Sir John *Enquiry into Bathes* 52
Fogel, Martin 152
Ford, Brian J. 'Leeuwenhoek specimens sent to Oldenburg', 57–8; 'Leeuwenhoekiana of C. Dobell' 58
fossils 36–7
Framlingham, Suffolk: Henry Sampson, intruded minister 1, 68
France 2, 35, 68
France, Joseph, ownership signature 145
Frati, C. *Bibliografia Malpighiana* 87
French Church of the Savoy, London 22
French translations 10–11, 13, 21–2, 47, 82, 88, 95, 129
'Fruits, Discourse' 25–6
Fry collection Herbals 92
Fulton, John F. xv, 18, 55, 83–5, 89, 95, 97, 136, 138
Funeral Sermon 72–3, 154

Gain, J., printer 127, 130
Galen 17
Galileo 54
Gallet, J.C. 164
Gennes, – de 164
German translation 47, 130
Gesner, Conrad 35, 38
gift inscriptions 76, 78, 100, 117, 145–6
Gilson, D.J. 92
ginseng root 36
Girin, Barthélemy, printer 11, 23, 97
Glisson, Francis 9, 12–3, 42–3
glossopetrae 36
Goddard, Jonathan 163
Gordon, A. 'Henry Sampson' 69
Goodall, Charles 32
Gough, Richard, books and bookplate 100, 117, 135, 137
Graves, John 163
Gray, P.H.K. *Catalogue of Mummies* 136

Greek: language 69; philosophy 8, 68; science 35
Green, Joseph Henry 145–6
Green, Thomas *Account of Virginia* 63
Greenland canoe 34
Gresham College 38
Grew, Elizabeth (Dodson) 3: Mary (Huetson) 167; Obadiah 1, 3, 5, 150
Guattoni, M. 164
guns 34
Gunther, A.E. 'The Royal Society and the British Museum' 39

Haak, Theodore 22, 33–4
hairs of plants 12, 26–7
Hale, Richard, bookplate 117
Hales, Stephen *Philosophical Experiments* 44
Hall, A.R. and M.B. *The Correspondence of Henry Oldenburg* xv, 2, 10, 60, 70, 147, 152
Halley, Edmond 54, 164
Hamersley, Mr, of Coventry 63
Harefinch, John, printer, 124–6, 128–9, 131, 133
Harley, Robert, Earl of Oxford 168
Harrison, J. R. *The Library of John Locke* 53, 118, 145; *The Library of Isaac Newton* 118
Harrys, Edw., ownership inscription 117
Harvey, William 41
Havers, Clopton 52
Hebrew language 55, 69, 72
Heffer publishers 95
Heidan, Abraham, Rector of Leiden University 5, 76
Held, Georg 152
Henrey, Blanche *British Botanical and Horticultural Literature* xv
Henshaw, Thomas 9, 116
Hevelius, Jan 54, 56, 152
Hickman, Spencer, printer 77–8
hippopotamus (*recte* rhinoceros) tooth, fossil 36–7
Hobbes, Thomas 54, 163, 167–8
Holford, Samuel, printer 120
Holles, Mr 46
Holt, Robert 65
Hooke, Robert 2, 7, 11–2, 18–9, 24, 26, 28, 32–4, 39, 49, 55–8, 158, 164; *see also* A.R. and M.B. Hall

Hooper, Richard, signature and book-label 92
Hoskin, Michael C. 21, 95
House of Lords 46–7
Howard, Charles 34, 163
    Henry, 6th Duke of Norfolk 36, 38
    Thomas, 14th Earl of Arundel 38
Howe, George 72; John *Discourse of future Blessedness* 68–9, 151
Howman, Dr 24, 158
Hubert (Hubbard), Robert 28, 35
Huetson, Mary, Mrs Grew 167
Huguenot refugees 14, 22
human anatomy 36, 42
Humboldt, Alexander von 16
humming-bird 60, 62–4, 148
Hunter, John 39–40
    Michael xvi, 2, 18, 25, 28, 50, 59, 66, 70, 152
    Richard, *Three hundred years of Psychiatry* 55
    William 105, 117
Hutchinson, William 105
Huygens, Christian 152
Hyde, Henry, 2nd Earl of Clarendon 157–8

Idea of a phytological (philosophical) History of Plants 8, 14–7, 25, 59, 85–8, 167
Imperial Academy (Academia Caesarea) 82
Indians *see* Amerindians
Indies, East and West 15, 34
Innes Smith, R.W. *English-speaking Students at Leiden* xvi, 1, 5, 69
insects 35
instruments, scientific 32–3, 38
Ireland 64
isenglas 42
Isham, Z., ownership signature 118
Italian translations 10, 48, 84–5, 132
Italy 2, 35, 68

James II 53
Jamieson, D.R. xvi
Jammes, Paul, bookseller 80
Jerusalem 54
Jesuits *see* Society of Jesus
Johnson, –, chymist 51; D.J. 46; E.A.J. *Predecessors of Adam Smith* 66; Samuel, *Dictionary* 11, 105
Johnson Reprint Corporation 105
Josten, C.H. *Elias Ashmole* 152

Journal des Scavans 58, 164
Jupiter, occultation 56

Keith, Sir Arthur (on Evelyn's 'Anatomical Tables') 39
Kempthorne, John 47, 123, 133
Kepler, Johann 60
Kettilby, Walter, printer 137–42
Keyes, Sir Geoffrey xiv, 12, 18, 24, 39, 57; *Bibliotheca Bibliographici* 78, 80, 83, 92, 117, 126–7, 129
Kildare, Earl and Countess of, *see* George Fitzgerlad and Jane Boyle
King, Edmund 42
Kircher, Athanasius 26, 38
Kittredge, George L. 67, 150

Laet, Jan de *Historia naturalis Brasiliae* 63
Lambert, Richard, bookseller 155
lamp-furnace, Hooke's 33
Lansdowne, Marquess, of, *see* Petty-Fitzmaurice
La Roque, J. B. de 58, 152
Latin drafts and versions 15, 17, 19, 47, 56–7, 85, 87, 89, 90, 165; Latin writings published 14, 42, 49, 57, 62, 135, 147, 165
Lawrence, Sir John 117
Lawson, John 135
Leadbetter, M.J. 61
leaves 12; 'Leaves, Discourse' 25–6
lectures to Royal Society 7, 17, 21, 23–6, 36, 40, 43, 97, 99, 105, 111–2
Ledbury, Herefordshire 21
Lee, Samuel 67, 150
Leeuwenhoek, Antoni van 33, 57–9, 96, 152, 164–5
Legati, Lorenzo *Museo Cospiano* 57, 164
Leibniz, Gottfried Wilhelm 152
Leiden University xvi, 1, 2, 5, 9–10, 21, 68–9, 76, 81–2
leopard 35, 41
Letters Patent of the States General of Holland 134
Le Vasseur, Louis 10–11, 13–4, 21–2, 79–81, 96, 152, 167
Lewis and Clark botanical Expedition 16
Lhwyd, Edward 109
libraries: *see* separate Index, following

light, theory of 34
Linnaeus, Carl von 16, 78, 117
lion 35, 41
De Liquore nervoso 5–6, 76
Lister, Martin 10, 21, 24, 57, 70–1, 78, 80, 92, 94, 100, 135, 153, 155, 164
'Lixivial Salts, Discourse' 25
loadstone *see* magnet
Locke, John 53, 87, 118, 145
London xi, 3, 5, 59, 68; Bishop of, *see* Henry Compton; *London Gazette* 52; London Wall 72
Lonicera periclymenum 12
Lorenzini, Stefano *Osservazioni intorno alle Torpedine* 59
Louis XIV 53, 80
Lovell, Robert *Panbotanologia . . . enchiridion botanicum* 13–4
Lower, Richard 44
Luctation see Experiments in Consort of Luctation
Lycidas (Milton) quoted 12

McAdoo, H.R. *The Spirit of Anglicanism* 53
Macalpine, Ida and Richard Hunter *Three hundred Years of Psychiatry* 55
Macdonnel, Randal (distilled salt-water) 47, 129
McKitterick, David 118
magnet (loadstone, terella) 33
maize 34
Malling, John, lens-grinder 33
Malloch, Archibald *Finch and Baines* 39
Malpighi, Marcello xv, 8–10, 25–6, 30, 51, 70, 87, 152; *Anatome Plantarum* 19, 23, 58; *De externo Tactus Organo* 61; *Correspondence see* H.B. Adelmann, editor
Malthus, Thomas, printer 118, 120
mammals, anatomy 35, 41, 111, 115–6
Mancetter, Warwicks; birthplace of Grew 1
Markgraf (Marggravius), Georg 63
marriage certificate: Grew/Huetson 167
Martin, John, printer 15, 17, 20, 86, 89, 91, 93–4
Mary II, Queen (Princess of Orange) 167
Massachusetts 35, 67, 72, 168
Massey, Dr 111
mastodon (*Tetrabelodon angustidens*) 37
Mathew, Richard: 'Mathew's pill' 18

Matthew, William, apothecary 121
Maule, Thomas 46, 131
Mauleverer, Thomas, ownership signature 100
measurement 43
medicinal plants 14, 27, 36, 112, 115
medicine, chemistry in 21
Mentzel, Christian 152
Merrett, Christopher 32, 116, 163, 165
Merrick, James, ownership signature 78
Mesmin, Guy 10–11, 21–2, 81, 96, 167
Metcalfe, C. R. 'Nehemiah Grew' xvi
Mexico: 'tigers' 35
Miall, L.C. The early Naturalists 17
Michallet, Estienne, printer 11, 23, 83, 95–6
microscope 7, 12, 19–20, 26, 33
Miege, Guy, translator 126, 129
Milford, Matthew 164
Miller, Genevieve 150
Millington, Sir Thomas 24–5, 32, 136
Mills, A. A. 'Newton's telescope' 33
Milton, John, Lycidas quoted 12
mind, theory of 55
mineral waters 49–52, 136–42
minerals, chemical experiments on 21
Miscellanea curiosa Academiae Naturae
   Curiosorum 11, 15, 17, 19, 81–2, 87, 90, 92
Mixture see Discourse concerning Mixture
mole, anatomy 41
Molyneux, Sir Thomas 152
Montague House (British Museum) 39
Montbéliard, P.G. de Les Oiseaux 64
Montesquieu, Charles de 80
Moore, John, Bishop of Ely: books at
   Cambridge 24, 100, 117, 120, 135–6, 145
Moore Smith, G.C., ed. The Letters of Dorothy
   Osborne 49
Moray, Sir Robert 163–165
'De morboso Liene' 60, 62, 147
morphology 27, 40
Morse, Thomas, ownership inscription 145
Mortimer, Cromwell 110–11
Motte le Vayer, F. de la 94
Moult, Francis, apothecary 50–51, 136–8,
   George 50–51
movement in plants 20
Moxon, Joseph Mechanic Exercises 163–4
mummy, Egyptian; – medicinal 36
Munificentia Regia 1715 bookplates, see John

Moore Bishop of Ely
Musaeum: Repository of the Royal Society
   xvi, xvii, 2, 8, 23–5, 27–43, 63–4, 68, 72,
   109–21, 155–61, 167, Plates, 1, 2, 7
museum catalogues consulted for Musaeum 38
muscle, colour of 42

naturalists in antiquity and Renaissance 35
Nazzari, Francesco 52
Needham, Joseph Clerks and Craftsmen in
   China and the West 60; –, Walter 18, 32, 43
nervous liquor 5–6, 76
Nevis 132
New England 63, 67, 71
New Experiments concerning Sea-water made fresh
   44–8, 132–4
New Repository 38
New South Wales 41
Newman, Hugh, printer 121
Newton, Sir Isaac 18, 21, 33, 38, 54, 56–7,
   110, 118, 159
Nigrisoli, F. M., translator 85
Noblett, W. 'William Curtis's botanical
   library' 105
nomenclature 11–15, 42
Norfolk, Duke of, see Henry Howard
North America 34, 41, 62–3, 65–8, 71 see also
   Index of Libraries
Northaw water 49
Notes and Records of the Royal Society 2, 33, 38,
   58, 110
'Number of Acres in England' 64–6, 148–9

'Observations touching the Nature of Snow'
   60, 147
'Observationes de morboso Liene' 60, 62, 147
Oglethorpe, Theophilus 46, 131
Old Jewry, London 72
Oldenburg, Henry xv, 2, 9, 10, 14, 28, 32, 36,
   55–8, 60, 68, 70, 96, 147, 152–3;
   Correspondence see A.R. and M.B. Hall
Orange, Princess of, see Mary II, Queen
Orléans, Philippe, Duc d': Regent of France
   80
Ormonde, Mary, Duchess (of 2nd Duke): her
   illness 51, 70
Ornstein, Martha The Role of scientific Societies
   in the seventeenth Century 32

Osborn, James Marshall: library at Yale 145
Osborne, Dorothy *The Letters to Sir William Temple* 49
Osler, Sir William 118; Library at Montreal, *see also* Index of Libraries
Oviedo, Juan Gonzales *Historia ... de las Indias orientales* 63
Oxford 3, 5; Physic Garden bookplate 100
Oxford, Earl of, *see* Robert Harley

Packard, Frances 'Henry Sampson' 69
Padua 39, 69
palaeontology 37
Pallas, Peter Simon 16
Palmer, Dudley: seven-shot gun 34
Palmer, Roger, Earl of Castlemaine 164
panther 35
papillae *see* skin-ridges
'Papers of Dr N. Grew' 167–8
Papin, Denis 90
Parini, Luigi, printer 84
Paris 13–4, 21, 35; physicians 167; *see also* Académie, and Index of Libraries
Parker, I. *Dissenting Academies* 3
 R., printer 134
Parkhurst, Thomas, printer 151
patents, patentees xvi, 46, 51, 73, 131, 134, 138
Pearson, John, surgeon 146
Pembroke College, Cambridge 1, 5, 8, 68; James Coates, college butler 168
Pepys, Samuel: books and bookplate 24, 89, 100, 117, 145; *see also* Pepysian Library in Index of Libraries under Magdalene College, Cambridge
Perceval, Spencer George 120
Perrault, Claude, anatomist 35
Perry, William *Bibliotheca Norfolciana* 38
Peter, Joseph *Truth in opposition to Falshood* 46, 49–52, 70, 137, 141–53
Petiver, James: extracts from *Musaeum* 109
Petty, Sir William 32, 49, 64, 89, 152
Petty-Fitzmaurice, William, 1st Marquess of Lansdowne 64, 148
Peyer, Johann Conrad 43
Philosophical Collections 56–8
Philosophical Transactions 2, 24, 32, 35, 39, 42, 44, 56–8, 60–66, 147–9, 163–5
phosphorus 33–4

photosynthesis 16
Physiologus legend 35
physiology of plants xi, 8
Piacenza, D. da 164
pith 7, 12, 16, 20
Pitt, Edmund 164
 Moses 59, 155
Plomer, H.R. *Dictionary of Printers* 155
Plot, Robert 152
polecat 41
pollen 12, 25
population of England and the Netherlands 64
Poplar, D., licenser 134
'Pores in the skin of hands and feet' 60–1, 147; pores in pith and wood 12
porpoise 15
portraits: Colwall xvii, 105, Plate 7; Grew xvii, 105, 143, Frontispiece
Portsmouth, Rhode Island 67, 150
Powle, Henry 163
Powys, Sir Thomas 47
Presbyterian Academy, Coventry 3; congregations xi, 53, 72; ministers, *see* Abraham Clifford, Obadiah Grew, Abraham Heidan, John Howe, Henry Sampson, John Shower
Prime, C.T., editor of Ray's *Flora* 9
printers: *see* Brewster, Bridge, Cademan, Chiswell, Clark, Darby, Dezallier, Elzevir, Gain, Girin, Harefinch, Hickman, Holford, Kettilby, Malthus, Martin, Michallet, Newman, Parini, Parker, Parkhurst, R-, Rawlins, Ray, Rogers, Roper, Roulland, vander Aa, Walford
priority in discovery 10, 23
Pritzel, G.A. *Thesaurus literaturae botanicae* 85
project for economic advance 64–6, 148–9
proposals (prospectus): *Anatomy of Plants* 23–4, 97–8; *Musaeum* 30, 72, 111–2
protozoa 59
psychiatry 55
publishers: *see* Heffer, Johnson Reprint Corporation, Rowman and Littlefield, Routledge and Kegan Paul, Wheldon and Wesley
Purkinje, Jan Evangelista (on finger-prints) 6

Queen of England *see* Anne; Mary II

Quendon, Essex: Abraham Clifford, intruded minister 5
questionnaires 67–8, 150, 167
Quinly, Jane *Catalogue of the Hunt Botanical Library* 80

R., J. printer 154
Racket Court, Fleet Street, 38, 72, 150
raingauge 33
Rastell, Thomas 165
Raven, C.E. *John Ray naturalist* 8, 18
Rawlins, William, printer 23, 93, 112
Ray, John, naturalist 8–9, 15, 18, 41, 51
 Historia plantarum 106; *The Wisdom of God in Creation* 53
Ray, John, printer in Dublin 127
Reisel, Solomon 152
remora, see ship-halting fish
Repository *see Musaeum*
Rhine at Schaffhausen: beavers 35
Rhinoceros antiquitatis 37
Rhode Island 67, 150
Risley, Derbyshire 1
Robinson, H.W., ed. *The Diary of Robert Hooke* 57
 F.J.G. *Book subscription Lists* 25
Rogers, W., printer 142
Rondelet, Guillaume 38
Roots see Anatomy and vegetation of Roots
Roper, A., printer 134
Roulland, L., printer 11, 79
Routledge and Kegan Paul, publishers 106
Rowman and Littlefield, publishers 106
Royal African Company 39
Royal College of Physicians, London 3, 14, 39, 42, 44, 46, 50, 68, 72, 132
Royal College of Surgeons of England 39
Royal Society xi, 2, 7–10, 14, 17–20, 22–25, 42, 49–51, 53, 55–66, 68, 70–1, 76, 78, 112, 116, 145, 155–6, 167; Archives 93, 109–11, 147, 152; Library ix, 38; Repository *see Musaeum; see also Philosophical Transactions*
ruminants 41
Rupert, H. R. H. Prince 33–4
Russell, Kenneth Fitzpatrick *British Anatomy* xvi, 112, 120–1
Russia 35
Ryan, M.J. 'Villalpando' 54

saffron-kiln 34
St Michael's Church, Coventry 1
Sakula, Alex xiv, xvi, 49–50, 61, 72
Salisbury, Bishop of, *see* Gilbert Burnet, Seth Ward
Salter, T. Bell, book-label 120
salts 25–6, 36, 44–9
Salviani, Ippolito
Sampson, Helen (*née* Vicars, later Grew) 1, 68
 Henry 1, 2, 8, 9, 24, 32, 35, 43, 57, 68–9, 151, 164
 William, father and son 1, 68
Sancroft, William, archbishop 24, 32
sap 12, 16, 20
Sarfati, apothecaries 51
Schaffhausen 35, 43, 69
Schoonebeck, A., engraver 82
Scott, Mr (voyage to India) 107
Seadgel, E.G., bookseller 154
Sea-water made fresh 44–8, 73, 122–34
Sedgwick, Thomas 62
seeds 15, 27, 99, 106–7; 'Discourse' 25
Settala, Lodovico 38
Severino, Marcantonio *Zootomia* 41
Sewall, Samuel 67, 150
sex in plants 9, 12, 25–6
sharks' teeth 36
Sharp, Thomas, archdeacon: bookplate 137
Sharpe, John, archbishop 53
Sharrock, Robert *The History of Vegetables* 12
sheep, stomach of 41
shells 37
Sherard, William: book collection 100, professorship of botany 117
ship-halter fish (*echeneis, remora*) 37
ships 65; sailing against tide 33
Shoe Lane, London 72
Shower, Sir Bartholomew 46, 73
 John 53, 154, 168; *Funeral Sermon* by 72–3
shrewmouse 42
Siamese drum 34
Sibbald, Sir Robert 152
silk-moth (*Bombyx*) 35
Simon, Ed. ownership signature 96
Simpson, A.D.C. 'Newton's telescope and the cataloguing of the R.S. Repository' 33, 38, 110
Singer, Charles *Science, Medicine and History*

150

skeletons, comparative measurements 36

skin, human, tanned 36; skin-ridges (papillae) 61

Slare, Frederick: preparation of phosphorus 33–4

Sloane, Sir Hans 3, 38–9, 49, 107–8, 152–3; herbarium 107; manuscripts 8, 15, 56, 58, 71, 85, 106–9, 150, 152, 167–8

Sluse, René François 152

Smith, Sir James Edward 78, 117

S., printer 50, 135, 140, 143

snails, hermaphrodite 25–6

snow, 'Observations' 147

Society of Apothecaries 50

Society of Jesus: College at Rome 26, 38; Library in Paris 164

'Solution of Salts in Water', Discourse 25

Somerset House, Royal Society moves to 39

Sotheby's sale of Royal Society books 38

Southwell, Sir Robert: crocodile skeleton given to Repository 39; P.R.S. 135

Spanish translation 47, 130

spectacle lenses 12

spiders 35

Spinoza, Benedict 53–4

spleen, disease of 62, 78, 147

Springell, F.C. Connoisseur diplomat (Lord Arundel) 38

squirrel, Virginian 41

Stanley, William, chaplain to Princess of Orange 167

States General of Holland and Zeeland 134

stems ('trunks') of plates 18–20

Steno, Nicolaus, identifies glossopetrae 36

stomachs see Comparative Anatomy of Stomachs and Guts

Stonestreet, Nicholas, ownership inscription 136

Streatham water 49

structure of animals 26, 40; of plants 20, 27

subscription publishing 23–5, 30, 32, 97, 112, 155–61, 167

substance of plants analysed 26

surveyor's wheel – 'waywiser' 33, 64

surveys of England and Ireland 64–6, 148–9

Sutton, Thomas, apothecary 107

Swab, Anton von 78, 92

Swammerdam, Jan 42, 152

Swift, Jonathan: Voyage to Laputa 65; on Conformity Bill 168

Switzerland 2, 68

Sydenham, Thomas 32

Sydow, C. O. van, Keeper of Manuscripts, Uppsala 78

Sylvius see Deleboë

Symonds, Nath. ownership inscription 145

Tabulae Harveianae 39

tastes of plants 25–6

Tavernier, J.B. Voyages reviewed 163

telescopes 33, 38, 110

Temple, Sir William 49

Tenison, Thomas, archbishop 53, 142

terella see magnet

terminology 11–15, 21, 42, 75, 88

Theobald, Mr 111

Theophrastus; on plants 8, on sex in plants 26

Thomas, Oldfield 'Zoology' in BM (NH) History of the collections 40

Thompson, D'A.W. On Growth and Form 60

Thomspon, Robert 152

Thoresby, Ralph 152

thorns of plants 12

Thornton, J.L. and R.I.J. Tully Scientific Books 100

Thruston, Malachi 165

tiger 35, 41

Tillotson, Thomas archbishop 24

'tooth, petrifyed, of a sea-animal' see mastodon

Torless, Richard 135

Tractatus de Salis cathartici amari Natura (Treatise of the Bitter purging Salt) 49–52, 135–42

Tradescant, John 41

Tramel, Thomas, apothecary 50–51

Trant, Peter 46, 131

Trapham, Thomas Health in Jamaica reviewed 164

True Catholic, anonymous pamphlet 168

Trunks see Comparative Anatomy of Trunks

Tully, R.I.J. see Thornton

Turnbull, H.W. ed. Newton's Correspondence 57

Turvey, P.J. 'Newton's telescope' 33

yson, Edward 24, 31, 61, 135, 164

ppsala University Library 78, 97
pton, Francis 72
vedale, Robert 107
alvulae conniventes 42
an der Aa, Pierre, printer 11, 82, 84
avilov, S.I. 18
egetation of Roots 14–7
elthuysen, Lambert van Tractatus de Liene 62, 78
enice printing 11, 81, 84
enn, J. and J.A. Alumni Cantabrigienses 5
ernatti, Sir Philiberto 163
illalpando, Juan Bautista 54
neyards in England 65
irginia 41, 63
oltaire, François Arouet de, quoted 80

agstaffe, William: author's gift of Musaeum 117
alcot, H. Treatise concerning . . . William Walcot 123
William 44, 46–7, 73, 122–3, 125, 134, 143
alford, B., printer 50, 135, 140, 143
aller, Erik, collection, see at Uppsala in Index of Libraries
allis, John 23, 33, 164
.J. Book subscription Lists 25
ard, Seth, Bishop of Salisbury 32
arrington Dispensary: sale of library 118
arwick Court, and Lane, London 30, 32, 72
arwickshire 1, 5, 76; see also Atherton, Coventry, Mancetter
ater absorption by seeds 27
ay-wiser (surveyor's wheel) 33, 64
eapons 34
eather-clock 33
eatherby, Dr 130
ebster, Charles 168
eit, Nehemiah 1

Wellcome Institute, London: Portrait catalogue quoted 143
Wellington, Mr. apothecary 21
Wepfer, Johann Jakob 35, 42, 69
Whalley, George 146
Wharton, Thomas Adenographia 42
Wheldon and Wesley, publishers 9
White, Robert, engraver xvii, 30, 120, 143, frontispiece and Plate 7
Wilkins, John, Bishop of Chester 2, 7, 9, 13, 28, 33, 64, 112
William III 53
Williams, David 146
    Moses, inventory of 'Repository' 110
Williamson, Sir Joseph, elected P.R.S. 56
Willis, Thomas, quoted 43, 54; Pharmaceutice rationalis 18
Willoughby, Sir Henry, Bt. of Risley 1
Willughby, Francis 43
Winchelsea, Earl of, see Charles Finch
'winders' (spiral climbing plants) 20
Winthrop, John, Governor of Conneticut 34, 165
Wittie, Robert 116
Wolf, Edwin, 2nd xiii, 125
Wood, Anthony, collection in Bodleian Library 112
    Neal 66
Woodcroft, Bennet Index of Patentees xvi, 44, 46
Woodward, John 110
Worm, Ole 35, 38, 109
Wren, Sir Christopher 24, 32–4, 38, 49, 56, 98, 152

York, Archbishop see John Sharpe
Yorkshire subscribers to Anatomy of Plants 24, 71, 155

Zeitlin and Ver Brugge, booksellers 105
Zirkle, Conway 26–7, 105
zoology in Musaeum 40

Index of Libraries
with selected copies of Grew's writings

England

Cambridge: Botanical School 100; Magdalene College, Pepysian Library 89, 100, 117, 145; Trinity College 100, 117, 118; University Library 78, 80, 83, 86, 92, 94, 100, 117, 120, 124, 126–30, 134–6, 145, 154

Eton College 100, 120

Leeds University: Brotherton Library 117, 120

London: British Library 55, 69, 76, 78, 80, 83–4, 86, 89, 91, 95–7, 100, 105, 112, 117–9, 121, 124, 135, 142, 145, 148, 150–51, 154: British Museum (Natural History) 78, 83, 85–6, 100, 117, 121; Chelsea Physic Garden 100; Guildhall 124; Linnean Society 78, 86, 92, 100, 117; Medical Society of London 141; Royal Botanic Gardens, Kew 78, 100; Royal College of Obstetricians and Gynaecologists 121; Royal College of Physicians 76, 78, 94, 100, 117, 145, 154; Royal College of Surgeons of England 78, 81, 90, 92, 94, 96, 100, 117, 135, 137, 145, 154; Royal Horticultural Society 78, 100; Royal Pharmaceutical Society 100, 117; Royal Society 76, 78, 93, 100, 109–11, 117, 120–21, 145, 147, 152; Royal Society of Medicine 78, 86, 92, 94, 100, 117, 137, 140; University of London 100; Wellcome Institute 78, 80, 86, 95–7, 100, 120–21, 124, 135 137, 141, 145

Manchester University 117, 129, 136

Norwich: John Innes Institute 86, 100

Oxford: Bodleian Library 21, 52, 78, 83, 89, 92, 94, 96, 100, 112, 117, 124–30, 135, 137, 145, 153–4; Christ Church 70, 124, 127, 130, 132, 137; Magdalen College 94, 137; Radcliffe Science Library 100; Taylor Institute 92

Reading University, Cole Library 78, 97, 100, 105, 117, 121, 145

Reigate, St Mary's Parochial Library 145–6

Ireland

Cashel Diocesan Library 127

Dublin: Marsh's Library 80; National Botanic Gardens 100; Royal College of Physicians 87, 117; Royal College of Surgeons 100; Trinity College 78, 80, 86, 100, 117, 121, 127, 135

Scotland

berdeen University 94

dinburgh: National Library of Scotland 117, 120, 145; Royal Botanic Gardens 78, 84, 86, 100; Royal College of Physicians 100, 117, 121, 137, 140, 145; University Library 86, 92, 100, 117, 120–21, 145

lasgow: Royal Faculty of Physicians and Surgeons 121; University Library 100, 117, 121

t Andrews University 120

Wales

angor: University College 100

New Zealand

Vellington: Turnbull Library 121

France

aris: Bibliothèque interuniversitaire de Médecine 80–1, 83, 97; Bibliothèque Nationale 78, 80, 83–4, 86, 89, 94, 97, 105, 117, 125, 127–9, 135, 145

Netherlands

.msterdam University 83, 105, 117, 135; Groningen University 78, 92; Leiden University 76, 83, 92, 117; Leiden, Boerhaave Museum 80, 105; Utrecht University 83, 95, 120; Wageningen Horticultural College 83, 105

Sweden

Jppsala University, Waller Collection 97

NORTH AMERICA

Canada

Montreal, McGill University, Osler Library 76, 86, 92, 105, 118

United States

California: Berkeley 97; Huntington 94, 127, 134; W.A. Clark 94, 120, 127, 130, 136, 139, 154

Connecticut: Yale: Historical Medical Library 81, 83, 85, 89, 95, 97, 136, 138; University Library 92, 94–5, 97, 121, 124–5, 127–8, 130, 145, 154

District of Columbia: Washington: Folger Library 86, 92, 125, 128, 136–7; Library of Congress 83, 86, 89, 92, 105, 125, 128; National Library of Medicine (Bethesda) 76, 78, 80–1, 97, 105, 121, 127, 135–6, 145; Smithsonian Institution 121

Kansas: University Medical Center 136, 138

Maryland, Baltimore: Institute of the History of Medicine 118, 145

Massachusetts: Boston and Cambridge: Harvard, Arnold Arboretum 86, 89, 92, 94; Countway Library of Medicine 85; Houghton Library 92, 121, 124–5, 127–8, 145, 154; Zoology Library 120; Massachusetts Historical Society 94, 137–8 Wellesley, Babson Institute 118

Michigan: University, Ann Arbor 86

Missouri: St Louis, Missouri Botanic Gardens 81, 87, 94

New Jersey: Princeton University 120

New York: N.Y. Botanic Gardens 81, 84, 86, 92, 121; Columbia University 121, 125–6, 129; N.Y. Academy of Medicine 92, 128, 136; N.Y. Public Library 121, 126, 129; Ithaca: Cornell University 97

North Carolina: University, Chapel Hill 81

Ohio: Oberlin College 92

Pennsylvania: Carlisle, Dickinson College 136: Philadelphia: College of Physicians 105; Jefferson College 81; Library Company 125, 128; University of Pennsylvania 86–7. Pittsburgh: Hunt Botanical Library 80, 105

Virginia: Williamsburg, William and Mary College 84

Wisconsin: University, Madison 89, 92, 97, 127–8